Analytical Chemistry and Microchemistry

nova
science publishers
www.novapublishers.com

DOI: https://doi.org/10.52305/OTNA9625

NOTICE TO THE READER

Library of Congress Cataloging-in-Publication Data

ISBN: 979-8-88697-059-3

Published by Nova Science Publishers, Inc. † New York

Contents

Preface

Pharmaceutical Chemistry is a scientific discipline in the chemical and pharmaceutical field involved in the design and development of pharmaceutical drugs and their analytical techniques. Analytical chemistry is a branch of Pharmaceutical Chemistry that specializes in the analysis of various substances. It involves the separation, identification, and the quantification of matter by using classical and modern methods involving the use of scientific instruments. This book is an attempt to expose the analytical techniques used in pharmaceutical chemistry, so that one can possess command in analytical techniques.

It gives me great pleasure to put in front of a large community of pharmaceutical sciences, my understated work on *Functional Analytical Techniques in Pharmaceutical Chemistry*, written in accordance with the latest need for the analysis of pharmaceuticals. My purpose in writing this book is to introduce the basic principles of the analytical methods in modern lines. I hope this book will be useful and meet the needs of readers at all levels.

Functional Analytical Techniques in Pharmaceutical Chemistry enables students, academic researchers, industrial pharmacists and analysts working in pharmaceutical sectors, forensic science and drug testing laboratories, and persons working in quality assurance and quality control departments to gain a basic knowledge of concepts, techniques and the use of analytical tools. Unique emphasis on practices in pharmaceutical laboratories, such as preparation of samples and techniques used for separation, provides an effective and efficient framework. Suitable for basic studies, this important text presents the most common analytical methods used in quantitative and quantitative chemical analysis.

I would like to thank *Dr. Bashir Ahmad Ch.* (Ex-Dean Faculty of Pharmacy, Bahauddin Zakariya University Multan, Pakistan) and *Dr.*

Muhammad Uzair (Dean Faculty of Pharmacy, Bahauddin Zakariya University Multan, Pakistan) for their kind supervision. I would also like to thank all my colleagues for their excellent guidance and support during this process and the coauthors, especially *Mr. Muhammad Rashad.* Without their cooperation I would not have been able to do this job.

I would like to thank the publisher of this book, *Nova Science Publishers,* for their cooperation and support. Debating topics with my friends and family was also beneficial to me. If I ever lose interest, it keeps me motivated. My parents deserve a special note of appreciation: your wise advice and kind words, as always, have been of great help to me.

I hope you enjoy your reading.

Dr. Malik Saadullah
PhD Pharmaceutical Chemistry
Government College University
Faisalabad-Pakistan

Acknowledgment

I would like to thank my parents and teachers, especially Dr. Bashir Ahmad Ch. (Ex-Dean Faculty of Pharmacy, Bahauddin Zakariya University Multan, Pakistan), Dr. Muhammad Uzair (Dean Faculty of Pharmacy, Bahauddin Zakariya University Multan, Pakistan) and Dr. Muhammad Asif for their motivation, guidance, and proofreading. I would also like to thank all my colleagues for their excellent guidance and support during this process. I would like to especially thank the co-authors Ms. Sania Arif, Mr. Muhammad Rashad, and the students for data collection and management. Without their cooperation, I would not have been able to do this job.

Chapter 1

Background

Malik Saadullah[1,*], Muhammad Asif[2], Sana Mehmood[1] and Muhammad Rashad[1]
[1]Department of Pharmaceutical Chemistry, Government College University, Faisalabad, Pakistan
[2]Faculty of Pharmacy, The Islamia University Bahawalpur, Pakistan

Abstract

The introduction of medicines has brought about a revolution in the realm of human health. This is only possible if the medicines are free of contaminants and supplied in the proper dosage. As a result, a variety of chemical and instrumental techniques have been developed that require regular estimates of medicines. As a result, these medicines must be identified and quantified at various phases of their manufacture, transit, and storage. Analytical apparatus and analytical methodologies play a key part in this process as well. As a result of this study, the relevance of analytical instruments and analytical methodologies in determining the quality of medicines is brought to light and underscored. For example, titrimetric and electrochemical approaches have been used to analyze medicines, as well as spectroscopic or spectrophotometric techniques.

Keywords: separation, quality control, extraction, chromatography, assay determination, validation, calibration, analytical techniques, instrumentation

---

*Corresponding Author's Email: Maliksaadullah@gcuf.edu.pk.

In: Functional Analytical Techniques in Pharmaceutical Chemistry
Editor: Malik Saadullah
ISBN: 979-8-88697-059-3

Introduction

In this book, you will learn about the concepts, instruments, and applications of modern chemical research. According to a nationwide assessment, the quantity of material allocated to each technology is dependent on its use in industry. It has been decided to leave out a few strategies that are not used often. The purpose of this presentation is to provide rapid concentration of information about techniques.

Analytical chemistry has competed with other techniques that are surrounded by other disciplines of technology. To be effective, an analyst must be able to sketch, conduct, and explain measurements in the context of the technical problem that he or she brought. The preference and consumption of appropriate method needs vast information of chemistry, at the same time, the knowledge and utilization of instrumentation is also important. In the end, the analyst must have wide knowledge about statistical tools and software to interpret his data to give proper information of the experiment by interpreting the results.

Qualitative analysis is when an examination is based on the recognition of one or more components of a sample. On the other hand, an experiment to investigate how much of a specific species is present in these components of samples is called quantitative analysis. Sometimes the knowledge regarding the positioning of atoms is necessary to establish the location of the individual atoms inside a molecule or compound for confirmation of a certain organic functional group. This type of examination is called structural analysis and is contemplated as an elaborate form of analysis. Species that are utilized either for quantitative or qualitative investigations are called *analytes*. There is much similarity between the methods and techniques used for quantitative and qualitative analysis of a species. For this purpose, samples are prepared both for quantitative and qualitative examination. Then quantification of this analyte is performed on the basis of some features related to that analyte.

Importance of Analytical Chemistry

Chemical analysis is an essential servant of modern technology and at the same time it depends upon modern technology for its applications to some extent. In fact, they are developed side by side. Quantitative chemistry and chemical analysis have played a crucial role for the production of chemicals from the beginning. Lavoisier's combustion investigations and Dalton's atomic theory were based on quantitative analytical confirmation. The transistor gives the latest example of an innovation that was impossible without the knowledge and development of tremendous and exact chemical analysis. That is especially fascinating as it explains the synergistic development that is regularly noticed in various fields. Having the support of the development of the transistor, the instrumentation for the analytical analysis now provides extraordinary uses for it. The value of analytical analysis is overestimated nowadays as we cannot exclude the importance of this technology. Some of the important applications of this technology are given below.

Fundamental Research

Identification constituents of any unknown system and development of new drugs is done by structural, qualitative and quantitative analysis. This is all done for the development of new drugs.

Product Development

Linkage between the chemical composition and physical properties of any system leads to new drug design. e.g., polymer composites.

Product Quality Control

To meet uniform drug quality either for new material or finished product, drugs are subject to chemical analysis. Constituent's quantities are kept as allowed by law.

Maintaining of Pollutants

Pollutants such as pesticides and metals are arrayed by sensitive chemical analysis and routine accurate analysis is done for its control.

Assay

For commercial significance, in an array of raw materials such as ores, a small amount is taken instead of a large concentration. Reliable analysis is done for this purpose.

Medical and Clinical Studies

Chemical analysis plays a vital role in the identification of different elements in body fluids e.g., how much sugar is in the urine.

Analytical Issues and Their Solutions

Both quantitative and qualitative analytical problems utilize the same solution on the basis of the same design. This is given below in seven different points.

Selection of Method

X-ray fluorescence spectrometry is selected for rapid quantitative analysis in trace elements. The right selection of a method plays an important role in resolving a problem.

Sampling

For reliable analysis, correct sampling is very important. When a sample that is truly representative of parameters is measured, it leads to accurate analysis.

Preliminary Sample Treatment

Sometimes a problem in finding a solution exists in the sample treatment like dying, ignition and dissolution. For solids, there is crushing, mixing and other prior requirements.

Separations

Influence from other constituents of samples leads to many problems. Techniques are valuable to separate such interferences. Avoidance of such interference leads to accurate analysis.

Method Validation

If results are meaningless, it is useless to carry out the analysis. Methods should be properly validated to make them valuable. A standard should be selected with a matrix similar to that of the sample and they must be re-analyzed on a regular basis for good analysis.

Assessment of Findings

Issues must be kept in mind during the evaluation of results through statistical analysis.

Types of Analytical Methods

They are classified as:

1. Classical analytical method
2. Instrumental analytical method

The *classical analytical method* involves gravimetry and titrimetry. It involves the connection between the physical measurement and analyte concentration. The *instrumental method* sometimes requires chemical manipulation before instrumental measurement.

Improvement that is sensitive, precise and accurate; needs the betterment of instruments and changes in techniques of analytical chemistry.

Laboratory robots that helps in the automation of an analysis plays a vital role. This makes the benchmark of the analysis rapid and efficient. *Microprocessors* in analytical instruments are a unique development. The microprocessor improves the performance of an instrument, maintains the laboratory information of an instrument and management system. The automatic recording of huge numbers of samples is made possible by implementation of this technology.

Computerized statistical analysis of data, chemometrics and expert systems give interactive computerized assistance and assessment in addressing analytical difficulties.

In an analytical approach, the demand of a long range of instruments are increasing day by day. Analysts place greater emphasis on the chemical form of an element in a sample rather than its degree of presence e.g., space probes, deep area studies. The list of the following definitions will help in understanding the concepts of analytical chemistry.

Accuracy

The degree of closeness of a measured value to the standard value through experimentation is known as accuracy.

Analyte

The component of a sample that will be examined quantitatively or qualitatively is an analyte.

Assay

A profoundly precise measurement, particularly of an essential constituent of a certain material. It is also used to check the purity of a material. e.g., it is used to check the purity of drug or constituents of a dosage form; such as constituents of a tablet.

Background

The data or calculations obtained from previously published sources other than the anlayte is called background. Background information can also be the role of different instruments, addition of reagents etc.

Blank

When a simulated matrix replaces the sample in an estimation and the conditions are identical to those under which a sample would be analysed.

Calibration

The instrument's response to analyte mass, volume, or concentration is compared to a standard known as amount of analyte. Between a standard and instrument response, a calibration curve is generated as a function of mass, volume, or concentration of analyte across a specified range. Calculate a calibration factor if a plot is linear. Calibration enables rapid computation of results. Volumetric apparatus are calibrated by weighing measured volume of water.

Concentration

The amount of specific substances present in given volume or mass. It is indicated as W/W, W/V and V/V.

Constituent

Components of a sample are called constituents.

Detection Limit

How much analyte can be determined with high confidence using certain methods.

Determination

Analyte quantification with a precision of greater than 10 percent of the quantity present.

Equivalent

There are certain analytical laboratories that still use this misleading phrase to describe the quantity of a material that may create, react with, or be indirectly equated with one mole (6.023 x 10), but it is no longer used.

Estimation

The quantity of an analyte contained in a sample is measured semi-quantitatively (with a precision of around 10 percent).

Interference

An impact that alters the behavior of the analyte in an analytical process.

Masking

Samples are treated with a masking agent to prevent interfering with the analyte's ability to detect other elements of the sample.

Matrix

In a sample, matrix refers to what's left behind after analyte has been accounted for.

Method

An analytical process is carried out using a method, which is a comprehensive explanation of all the instructions that are used.

Variability

When a measurement or outcome has a random or undetermined error, it is said to be variable. Standard deviation or relative deviation are common terms used to describe it.

Primary Standard

When it comes to main standards, these are substances with perfect stability and purity that may be used as a benchmark for the testing of other samples.

Procedure

Description of all steps in an analysis.

Reagent

In an analytical technique, the chemical moiety is primarily utilised to generate a specific reaction.

Sample

Analytical methods are used on samples, which are substances, or portions of substances, whose analytical properties are unknown.

Sensitivity

Analytical sensitivity refers to the capacity of an analytical sample that responds to minor changes in the analyte. If the calibration curve is linear, it is equal to the slope of the curve.

Standardization

When the concentration of an analyte or reaction mixture can be determined from its reaction mixture, it is termed standardizing.

Standard

When it comes to standardization, the term standard refers to an unadulterated material that has the ability to react in a stoichiometric manner with a reagent or analyte.

Standard Addition

It is a technique that measures an analyte reaction before and after its addition to a sample combination. As a result of this approach, all analyte computations are performed in the same matrix, which removes any interference effects caused by differences in the sample or standard composition.

Technique

Technique is the underlying concept of a collection of procedures.

Validation of Methods

Validation of methods via analyzing standards is primarily used to determine whether the results yielded by a particular analytical method are accurate or not. The validating of results by comparing them to established analyte and matrix standards is necessary in order to ensure that the results are as accurate as possible. Our approach is accurate and based on the recognized values for these verified standards after comprehensive study and by utilizing diverse ways.

Conclusion

The term "to analyse" etymologically involves to examine, study, learn, work out, weigh, etc. Natural and artificial materials can be characterized both qualitatively (what's there) and numerically using analytical approaches (how much is present). This chapter deals with the basis of analytical chemistry including the definition of analytical chemistry, importance of analytical chemistry, analytical issues and their solutions and classification of the analytical methods. In practice, separation, identification and quantification may constitute the entire analysis or be combined with other analytical techniques.

Analytical chemistry consists of classical and instrumental methods. The classical analytical method involves gravimetry and titrimetry. Instrumental analytical techniques may be used to separate samples using chromatography and electrophoresis. Analytical chemistry is also focused on improvements in experimental design, and chemometrics as well as on the creation of new measurement tools.

Conflict of Interest

None to Declare

Consent For Publication

None to Declare

Acknowledgment

None to Declare

References

Frenzel, W. (2002). Book review: Principles and practice of analytical chemistry. 5th edition. By F. W. Fifield and D. Kealey. Angewandte Chemie International Edition, 41(7), 1249–1250. https://doi.org/10.1002/1521-3773(20020402)41:7< 1249:AID-ANIE1249>3.0.CO;2-1.

Chapter 2

Separation Methods and Quality Control

Muhammad Uzair[1], Malik Saadullah[2,*], Eman Y. Santali[3], Ruby Bashir[2], Akhtar Rasool[3] and Shahid Shah[4]

[1]Dean, Faculty of Pharmacy, Bahauddin Zakariya University, Multan, Pakistan
[2]Department of Pharmaceutical Chemistry, Government College University, Faisalabad, Pakistan
[3]Department of Pharmaceutical Chemistry, College of Pharmacy, Taif University, Saudi Arabia
[4]Faculty of Pharmacy, Government College University, Faisalabad, Pakistan

Abstract

To separate a mixture of chemicals into two or more different products or product mixes, a separation process is used. A number of components are added to the final product. It is possible to get pure components by using advanced procedures and well-designed separation methods. Chemical characteristics such as boiling point, distribution coefficient, or physical qualities such as size, shape, mass, and density might be used for separation.

Keywords: quality control, extraction, mixture, chromatography, liquid-liquid extraction (LLE), extractive distillation and solid-liquid extraction (SLE), stable-phase microextraction (SPME), supercritical fluid extraction (SFE), dispersive liquid–liquid microextraction (DLLME) and microemulsion (ME).

---

*Corresponding author's email: Maliksaadullah@gcuf.edu.pk.

In: Functional Analytical Techniques in Pharmaceutical Chemistry
Editor: Malik Saadullah
ISBN: 979-8-88697-059-3

Introduction

Extraction is the first stage in separating the intended result from the raw material, whether it is a natural product, biological sample, or pharmacological complex. During extraction, one or more components are selectively removed from a mixture, using a liquid immiscible solvent. Component solubility in the appropriate phase determines the amount of feed components that may be transferred to the solvent. The components to be separated are enhanced (extract phase) and depleted (raffinate phase) in the extraction stage.

Liquid-liquid extraction is the most commonly utilized. When used in wastewater treatment, it removes heavy metals and acids from the water and aromatic chemicals from hydrocarbon mixes. Extraction comprises liquid-liquid extraction (LLE), extractive distillation and solid liquid extraction (SLE). LLE is a commonly used extraction method.

Extraction also includes the separation of phytochemicals from dried plant material. After the solvent diffuses into a solid matrix, the solute dissolves in the solvent, and subsequently it diffuses out of the solid bed.

Extraction agents are selected based on an analyte composition. When it comes to extracting solvents, Hildebrand solubility criteria categorizes them in ascending order of polarity.

Ethyl acetate < Acetone < Ethanol < Methanol < Acetone: Water < Ethanol: Water < Methanol: Water < Water

Maceration, percolation, and soxhalation are all methods. After the extractive solvent has been applied to it, the extractive substance is put in a thimble. Then it condenses and runs back through the sample. Eventually, it becomes more focused. Rotary evaporation is then used to remove the extracting solvent from the soxhalation concentration.

Types of Extraction

1. Liquid liquid extraction
2. Liquid solid extraction
3. Super critical fluid extraction

Liquid-Liquid Extraction

One of the most essential processes of separation is liquid-liquid extraction. In addition, it is a more crucial stage in the preparation of samples than advanced methods. According to its distribution coefficient, analyte separates into two immiscible phases (illustrated in Figure 1). Using organic and aqueous solvents, this process is known as liquid-liquid extraction (LLE). Sometimes, water-soluble contaminants are separated from an organic phase using LLE.

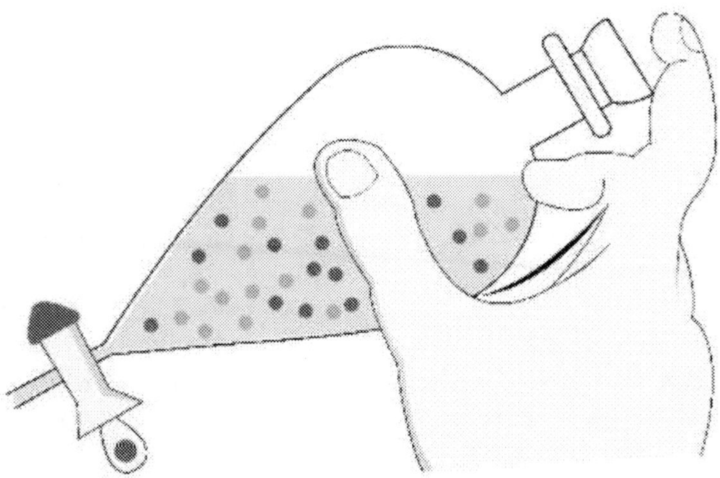

Figure 1. Liquid liquid extraction.

Extractors come in a wide variety depending on the type of analyte. Among the most popular solvents for drug and phytochemical extraction, there are cyclohexane, acetone, dichloromethane, chloroform, and water. When it comes to non-polar chemicals, hexane and cyclohexane are employed, whereas dichloromethane and chloroform are used for molecules that are moderately polar. The Soxhlet technique is used to separate lipidic fractions. Impurities and other interfering substances can be removed by employing these procedures. Extraction should be carried out using the most appropriate solvents, optimal temperature and pH conditions. It is possible to modify the extraction efficiency by altering the ionic strength of aqueous phase. When it comes to extracting medicines such as triazoles, amides, carbamates, and imidazoles, ethylacetate CH3COOC2H5 and dichloromethane CH2Cl2 are the most commonly used solvents. Prior to their extraction, some medicines, such as chlorophenoxy acids need to be derivatized. Dimethyl sulphate is used to derivatize chlorophenoxy acidic herbicides, which are subsequently extracted using hexane. An easy-to-use tool for this purpose would be the separation funnel, which has been used traditionally for this purpose.

Types of Liquid–Liquid Extraction (LLE)

Following are the types of liquid-liquid extraction.

Dispersive Liquid–Liquid Microextraction (DLLME)

In the aqueous phase, this method is utilized to extract tiny amounts of chemical molecules. An extraction solvent like tetrachloroethylene C2Cl4 and a disperser solvent like acetone are added to the water solution to achieve this effect. Afterwards, the solution is centrifuged to separate the organic and aqueous components. Organochloride and organophosphorus (pesticides) as well as substituted benzene compounds may be removed from water by using this method.

Direct Organic Extraction

Toluene, benzene, and xylene are organic solvents that are used to mix partly organic soluble materials. The organic soluble components will dissolve into the solvent and will subsequently be separated by use of a separating funnel. Phosphopeptide and phosphoprotein are separated using this extraction method.

Extraction Without Chemical Change

Solvent extraction is a simple method. In a closed container, two immiscible liquids are tossed together. Non-polar solutes will migrate to organic solvents and the polar ones are dissolved into them.

Ion Exchange Mechanism

It is an ion trading system. To maintain the charge balance, when one ion is transported from the aqueous phase to the organic portion, another ion is transferred through a different channel and this additional ion is a hydrogen ion. In an ion exchange process, the distribution ratio is a function of pH. Tert-butylbenzene and terpyridine, for example are used to extract americium.

Ion Pair Extraction

It is a lot more feasible to extract a metal by using a counter ion. If the nitrate content is high, it is possible to extract americium as a lipophilic quaternary ammonium salt from the mixture. This is an example of a section transfer catalyst. In an ion-transfer species, each time an ion interacts and creates another one, it is then transported back to the aqueous area for further processing.

Advantages of LLE

1. Wide variety of solvents with varying solubility.
2. Simple instruments.
3. Easy procedure.

Disadvantages of LLE

1. A major drawback in the use is the evaporation and disposal of organic solvents.
2. Although the glassware used is simple, its cleaning is difficult and it should be stored under rigorous conditions.
3. Some practical problems are involved i.e., formation of emulsion that are difficult to break once formed and it affects the efficiency of process.
4. Exhibits low performance when it comes to separation of polar compounds.
5. This is a good separation method for pre-treatment of sample before the use of sophisticated techniques.

Solid-Liquid Extraction

Solid phase extraction is another name for it (SPE). In a concept of affinity, a liquid (mobile phase) dissolves the solute and passes through a solid. As a result of its physical and chemical characteristics, analyte divides into two phases. For the purification and isolation of desired compounds, SPE is utilized. Because of the solute affinity for both phases, it separates into desired and unwanted components. In stationary phase, the components can be gathered individually. An extractive technique, SPE is also known as a chromatographic process.

In essence, SPE works in a similar manner to liquid-liquid extraction. Using one of these processes, dissolved matter is distributed between two stages. Analyte is dispersed between liquid (sample medium) and solid (sorbent) phases in this method rather than between two liquid phases that cannot be combined as in liquid–liquid extraction. It is possible to enrich and purify analytes on solid adsorbents by adsorbing them from the solution.

The solid phase components can be eliminated by eluting a solvent having a higher affinity for the analyte. Adsorbent material is placed in a polypropylene cartridge between two flitted discs and liquid phases are sucked or pumped through the cartridge.

We have a range of stationary phase materials that can be used to extract analytes from solid phase cartridges and discs, each of which has various chemical characteristics that can separate analytes. For example, most of these stationary phases are composed of silicon and some kind of functional group. Hydrocarbon chains of varying length (for reversed phase SPE), quaternary ammonium or amino groups (for anion exchange), and sulfonic acid or carboxyl groups are only a few of the functional groups to consider (for cation exchange). The classification of solid phase extraction is explained in Figure 2.2.

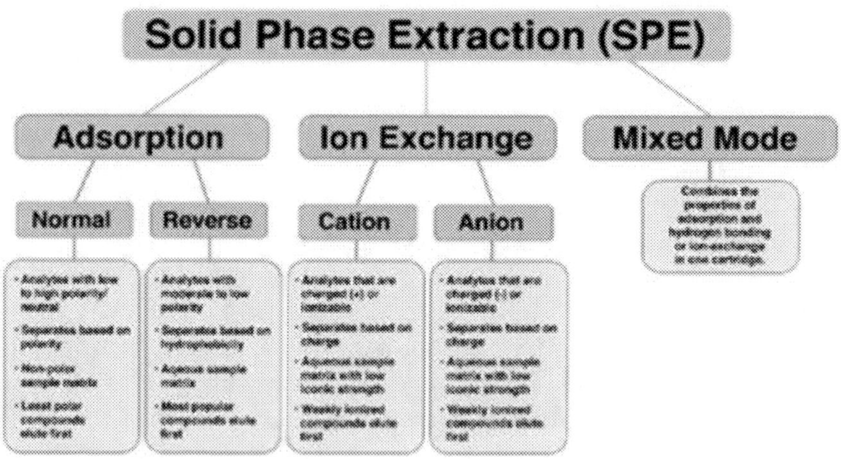

Figure 2. Classification of solid phase extraction.

Types of Solid Phase Extraction (SPE)

SPE are classified into following types.

Normal Phase SPE

This is the normal phase. It is utilized for analytes with a low-to-high or neutral polarity. The cartridge includes silica; a polar adsorbent. When the components are separated, the least polar ones will come out first. In most cases, the sample is placed in a non-polar matrix.

A Solid Phase Extraction Involves Basic Five Steps

In the first stage, a slightly polar or non-polar solvent is used to equilibrate the cartridge which allows it to penetrate the bound phase. After that, an aqueous buffer with the same composition as the sample is fed through the column to wet the silica surface. The sample is introduced to the cartridge, polar components or polar analytes in the sample are retained on the polar sorbent while the sample traverses through the stationary phase. The solvent and other non-polar contaminants flow through the cartridge. After the sample is added, the cartridge is cleaned with a non-polar solvent to eliminate any remaining contaminants. At the end, the analyte is eluted with a polar solvent or a buffer of an appropriate pH.

Reverse Phase SPE

The polarity of SPE analytes is used to separate them from each other in the process. Compared to the mobile phase, the stationary phase has a higher degree of polarity. In most cases, hydrophobic carbon chains are used to derivatize silica. These reverse phase cartridges are designed to hold analytes with low to medium polarity. When the stationary phase bed is wetted with a non-polar mobile phase, the analytes that have been kept on it can be separated. For the stationary phase, C-8 or C-18 chains are bound to silica. Non-polar to mild polar molecules are separated by hydrophobic contact.

Ion Exchange SPE

Analytes are separated using an ion exchange sorbent material based on electrostatic interactions between the analyte and functional groups on the stationary phase. It is vital to note that both the stationary phase and sample must be charged for ion exchange to take place, therefore pH is crucial.

Anion Exchange SPE

Negatively charged exchange solvents are derivatized with a positively charged solid phase to form anion. This occurs when negatively charged anions in the sorbent interact with positively charged groups on the stationary phase. There are sorbents that are very effective in anion exchange, such as those that have quaternary ammonium groups which maintain their positive charge in aqueous solutions. However, amine groups are charged when the pH is somewhat basic, about 9 in weak anion exchange sorbents. Highly acidic impurities are retained and checked for interference with the analyte of interest using anion exchange sorbents having strong anion exchange properties. A mild anion exchange cartridge should be used to recover a strong acid. A solvent is used to neutralize the charge of either an analyte or the stationary phase or both, in order to elute the analyte. Electrostatic contact between the analyte and stationary phase is eliminated after the charge is neutralized, allowing the analyte to be released from the cartridge.

Cation Exchange SPE

Positively charged ions are held in place by a stationary phase. Drugs are generally of a basic nature, with the exception of a few. If you use basic medicines in an acidic environment, you will get positive charges. Sulfonic acid groups are negatively charged while in aqueous solution, making it a strong cation exchange sorbent. Some weak cation exchange sorbents include aliphatic carboxylic acids which are charged when the pH is above around 5. A weak cation exchange cartridge is used to retrieve a strong base. This solution neutralizes any electrostatic interactions between the analyte and the stationary phase, allowing for easy extraction of the analyte.

Solid Phase Microextraction

A fibre coated with an extracting section which may be liquid (polymer) or stable (sorbent) is used in stable phase microextraction to extract a variety of analytes (both hazardous and non-unstable) from diverse media which can be liquid or fuel segment. During equilibrium or short-term pre-equilibrium, with the help of convection or agitation, the amount of analyte extracted by the fibre is proportional to its awareness within the pattern.

Supercritical Fluid Extraction

Isolation by supercritical fluid extraction (SFE) involves employing supercritical fluids as the extraction solvent in order to separate a substance (extractant) from another (matrix).

A stable matrix is usually used for extraction. For analytical purposes, SFE can be used to prepare samples. On a wider scale, SFE can be used to remove undesirable fabric from a product (e.g., decaffeination) or collect the desired product (e.g., important oils). SFE involves two key steps: the transportation of solid particles to the surface (via diffusion or any other method) and the dissolving of the solid particles in the supercritical fluid.

Since CO2 fluid in a supercritical extraction has a large density and dielectric regular and is highly solubilized in substances. The supercritical extraction device uses CO2 as a solvent. Supercritical CO2 extraction is particularly ideal for the extraction of high-boiling, fat-soluble, and heat-sensitive compounds as well as for the high-quality separation of different additives like supercritical distillation. To extract and purify numerous items such as organic commodities, food, and medicine, carbon dioxide is employed as a solvent.

Process of Super Critical Extraction

Supercritical extraction begins with a subcooling of the carbon dioxide before it enters the pump thus preventing cavitation and other problems. Once the pressurized CO2 is heated to a higher temperature, the extractor is kept at the proper operating temperature by using steam or hot water to keep it warm. Carbon dioxide (CO2) travels through the feedstock and into the extractor where its soluble components are extracted. This is followed by the solvent leaving with the extract and going to the separator (at atmospheric strain). Due to the depressurization of the separation vessel the extract is triggered. The process of super critical extraction is illustrated in Figure 3.

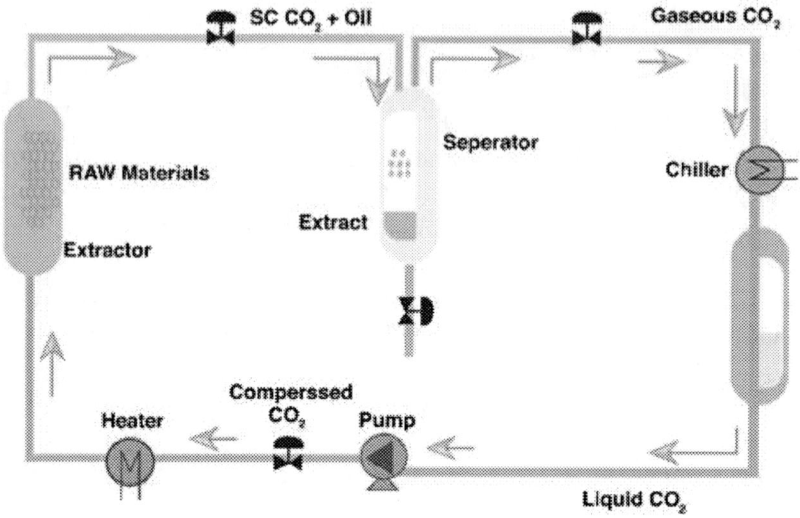

Figure 3. Process of super critical extraction.

Factors Affecting Extraction Efficiency

There will be greater separation if you use finer sized particles. The solubility and diffusion of the desired solutes into the liquids improves as the temperature rises. Increased contact time improves extraction until equilibrium is reached between the two phases.

While these factors affect extraction to a certain extent, also shows adverse effects upon completion of operation.

Quality Control Separation

There's no way to know for sure unless you go through the quality assurance procedure but the method and the analytical equipment have been proven to produce correct findings. These tests should be included as part of the quality assurance approach as they give information about typical performance of methods. Liquid chromatography has been linked to helpful methods that improve the overall quality of analytical findings according to recent research.

Chromatographic Fingerprinting

Medicines, especially of plant origin, may contain a lot of compounds. Despite the fact that many of these chemicals are present in low concentrations, they may be critical to the quality, effectiveness, and safety of herbal medications. They have a synergistic impact because of the various components that make up these herbal medicines, most of which have medicinal benefits. A variety of strong separation methods combined with appropriate detection instruments to serve as significant tools for the fingerprinting of these drugs. Chromatography is the most effective method for separating phytoconstituents from one another. IR, PDA, ECD, and FD detectors combined with the latest chromatographic methods provide an effective instrument for creating a profile of these substances. Botanical fingerprinting uses phytochemicals (herbal fingerprinting).

The development of herbal fingerprints based on biomarkers and is crucial for quality control of phytochemicals. As a result, it will be easier to identify the different types of chemicals in a sample combination. In this way, the action of a particular class of chemicals may be better understood. For quality control, scientists are now creating these herbal fingerprints. As an herbal medication, one of the biggest challenges is to isolate the required phytoconstituents from the rest of the plant extracts.

Thin Layer Chromatography

A mixture's non-volatile components can be separated using thin layer chromatography. On a sheet of glass, plastic, or aluminium foil coated with a thin layer of stationary phase (alumina, silica, or cellulosic), the entire experiment is carried out. Separation happens based on relative affinity for both phases when the mixture components move through the stationary phase. Along the vertical axis, separated components show as tiny patches of color. The retardation factor is used to determine quality of separation. The ratio of distance travelled by solute and solvent is called the retardation factor. Temperature, quantity of analyte spotted and solvent system are all factors that influence the separation process.

Assimilation of a chemical into stationary and mobile phases is based on its relative affinity. The component that has a higher affinity for stationary phase tends to travel slowly and the spot will appear around the base line of the component's trajectory. It would appear in the middle of the TLC plate, whereas components having a higher affinity for the mobile phase would move faster and separate more slowly at the solvent front. Separated fractions can then be analyzed using appropriate detection methods.

Advantages of TLC

1. Easy to use
2. Fast screening of samples
3. Cheap and readily available
4. Variety of stationary phases available commercially
5. Operational parameters such as plate development and derivatization can be optimized quite flexibly.

Disadvantages of TLC

1. Low reproductibility.
2. Resolution is poor when it comes to closely related analytes.
3. Sample to be anaylzed should be available in high concentration thus, less sensitivity.
4. TLC has semi quantitative nature more important for qualitative work.
5. Some factors are always hard to control like saturation in the chromatographic tank, sample spotting, unstable coloration when using chromogenic parameters of detection.

Thin layer chromatography, on the other hand, provides a benefit from recent advances in separation science. Stationary phases are now being used to improve resolution, repeatability and sensitivity of the measurements. When it comes to document recovery, the use of digital scanning machines and software has proven effective. There is a significant improvement in the use of TLC for complete separation, identification and assessment of phytoconstituent products.

High-Performance Thin Layer Chromatography (HPTLC)

Due to the development of high-performance thin layer chromatography which automates several stages such as the pressured passage of mobile phase across the TLC plates, the usage of traditional TLC is beneficial. Aside from that, the stationary phase with tiny particle sizes enhanced the resolution.

Microemulsion-Thin Layer Chromatography (ME-TLC)

Microemulsion plant mixtures are separated using ME-TLC as mobile phase. SDS is frequently employed as a surfactant in oil-in-water and water-in-oil emulsions and has demonstrated to be a superior surfactant for enhanced separation of different organic molecules according to research. ME-TLC usage boosted the separation efficiency, lowered detection limits, and improved sensitivity.

Densitometric-Thin Layer Chromatography (Densitometric TLC)

When using HPTLC scanners for qualitative assessments, the most important step is to make sure that the results are accurate. Digital data may be generated from TLC plates by using densitometric measurements. Other modifications rely on this information.

Two-Dimensional Thin Layer Chromatography

There are two separation methods performed on each sample. The separations are sequential, but they are not identical. Separation capability therefore increases.

Graft Thin Layer Chromatography

As a result of this approach, the separation takes place twice. There are two types of stationary phases that are employed. Using TLC, a partial separation is obtained before the same sample is moved to a different stationary phase and developed in a new dimension. Graft TLC is also called Coupled Layer Planar TLC (illustrated in Figure 4).

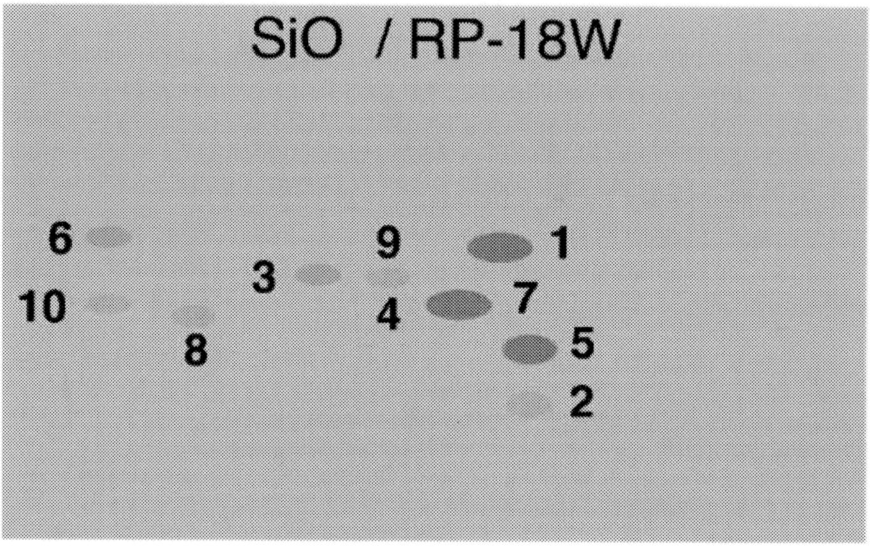

Figure 4. Representation of Graft TLC.

HPLC: High Performance Liquid Chromatography

It is the most widely used chromatographic method in the world today. It is a totally automated approach with excellent resolution, sensitivity, and selectivity, and it is easy to learn. Thanks to software, the entire process was simplified. Because pressure assisted separation can now be combined with sophisticated detection technologies and this technology is being used in larger operations.

Various substances can be separated using liquid chromatography. Unlike TLC, separation occurs in a column rather than on a flat surface like in TLC. The mobile phase runs through this packed bed when it is under pressure. In order to achieve greater separation, both high pressure and tiny particles of stationary phase are important factors to consider.

Analyte disperses itself between mobile and stationary phases depending on the situation. The on-column time of a molecule is determined based on the type and size of the analyte. HPLC system is diagrammatically shown in Figure 5, and it contains the following modules:

1. Solvent reservoir
2. Pump
3. Column
4. Detector
5. Data processing unit

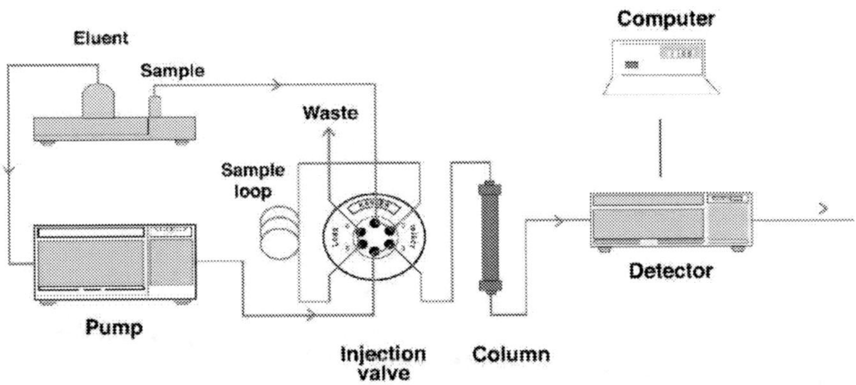

Figure 5. Schematic layout of HPLC system.

High-pressure solvent is pumped throughout the system at a consistent pace by a pump. As a result, it is essential to provide the detector with a pulseless and continuous flow of air.

UV, PDA, ELSD and CL can be used in conjunction with HPLC to identify phytochemical fingerprints. UV and diode array detector DAD are used for UV-absorbing compounds, whereas evaporative light scattering detectors and chemiluminescence detectors are utilized for non-UV-absorbing chemicals. Separated components can be identified using a mass spectrometer (MS). Nuclear magnetic resonance is utilized for metabolic research.

The choice of stationary phase is the most essential element for successful separation and fingerprint formation. This type of column is commonly used to separate phytochemicals. The particle size of the stationary phase has a major impact on the separation. Separation efficiency is improved by using smaller particles, however this also raises column back pressure. It has been shown that when particle size is halved, the pressure increases by four times. Assemblies for HPLC systems are intended to withstand the highest amount of back pressure possible. Up to 400 bar is supported with the current instrument. In order to obtain high separation efficiency in a short period of time, it is necessary to increase flow rate and decrease the particle size.

Disadvantages of HPLC

1. Expensive machinery.
2. Use of large volumes of environmental unfriendly chemicals.
3. Undetected co-eluents may affect separation efficiency.
4. Vulnerability of silica columns to withstand pH below 2 and above 9.

Ultra-High Performance Liquid Chromatography UHPLC

Separation methods for quality control of medicines have been continuously improved throughout the years to provide the highest possible quality. HPLC has evolved in recent years to be able to withstand higher back pressure. UHPLC can tolerate pressures of up to 400 bar (barometric pressure). Incorporating a very complex pumping mechanism, the device is only a modest hardware modification of standard HPLC. Some people refer to Ultra High Performance Liquid Chromatography as Rapid Resolution Liquid Chromatography. A stationary phase column with particle size smaller than 2 microns is used. This is due to the fact that smaller particles allow for better separations to take place. Reduced operating time and less solvent use can be attributed to the use of short columns. UHPLC requires customized detectors with tiny flow cell volumes and a significantly quicker acquisition rate than standard HPLC systems in order to obtain optimal responsiveness with narrow peaks through rapid analysis. Using ultra-high-performance liquid chromatography (UHPLC), numerous herbal fingerprints were discovered. Danggui Buxue Tang samples were subjected to UHPLC-DADTOF/MS analysis, allowing the identification of 39 components. A reduction in analysis time as well as an increase in selectivity were found.

Hydrophilic Interaction Liquid Chromatography

(HILIC) is liquid chromatography that employs either silica or silica-polymer-bonded-solid-layer stationary phases in conjunction with a mobile phase containing water as the major component of about 2.5 percent by volume in combination with less-polar solvents like acetonitrile and methanol water or aqueous buffers are employed with 5 to 45 percent ratio. Alpert created the term "HILIC" in 1990. There are several advantages of using it instead of conventional phase chromatography. Polar compounds can be separated by using polar stationary phase. Separation is based on the idea of partition. There is a hydrophilic layer that remains on the stationary phase and a mobile phase that is considerably less hydrophilic but still contains water up to 40 percent combined with other solvents. Combining the fingerprints of Ligusticum chuanxiong and Ganoderma species enhanced the quality control of the chemical compounds. Using chloroform as a mobile phase, Martin and Synge devised a technique for separating amino acids on silica columns in 1941. It was claimed that water in the mobile phase was separated from a hydrophilic layer on silica. Hydrophilic and charged molecules are retained preferentially in HILIC, which is the exact opposite of reverse phase chromatography. For example, toluene is substantially retained on C8 and C18 reverse phase columns, whereas it is eluted from HILIC empty volume. However, the HILIC separation of Uracil which is employed as a void volume marker in reverse phase column is excellent.

Gas Chromatography

Separation and characterizing volatile substances is done with the help of gas chromatography. With its high separation efficiency and sensitive detection mechanism, GC is a valuable instrument in the analytical process. Analyzing medicines, conducting research and testing natural products rely on gas chromatography. For the identification of compounds, GC combined with a mass spectrometer is a useful tool. The analyte must be volatile and have a molecular weight of less than 1250 Dalton in order for the analysis to be effective. When used with GC, sensitive detectors such as FID and MS may identify complex biological samples, herbal product and essential oils. Due to recent advances in GC-MS technology, it is now possible to analyze essential oils in as little as 40 to 100 seconds. When using GC, the detection thresholds are lowered. The separation efficiency is enhanced by the use of a micro bore capillary column, decreased stationary phase film thickness and fast temperature programming. Mega bore analytical columns resulted in a reduction of separation efficiency but an increase in analytical speed.

Miniaturized Separation Techniques

Due to their various benefits over traditional separation methods, miniaturized separation technologies are much sought after. Miniaturized methods are characterized by reduced analytical time, reduced usage of ecologically harmful chemicals and excellent sensitivity. As a result, the ability to inject huge volume of sample is severely limited. A pre-concentration process is used to circumvent this restriction which might be a hindrance in its frequent use. Columns and chips might be used in miniaturized methods. In order to circumvent this restriction, chip devices can be used to measure tiny quantities of analyte. In addition to liquid chromatography, a chip device may be used to perform gel chromatography and micellar electrokinetic chromatography.

Separation Techniques Isocratic versus Gradient Elution

Mobile phase can be pumped through a column using an elution process. While in the isocratic technique, the composition of the mobile phase does not vary but it varies in a gradient method. When constructing a separation, the isocratic approach should be the initial choice because it is the easiest. In most cases, eluent gradients are created by combining the pressure flow from two pumps and varying their separate flow rates with an electronic controller or data system while holding the total flow rate constant. Analytes are derivatized by chemical reactions that modify their molecular structure to enhance detection. Typically, HPLC does not need the derivation of drugs in order to get good chromatographic results. For underivatized chemicals, derivatization is utilized to improve sensitivity and selectivity of detection when existing detectors are not sufficient.

Conclusion

There are several different ways for isolating and fractionating chemicals from a sample mixture in order to gain information about their structure or characteristics. Liquid-liquid, liquid-solid and supercritical fluid extraction techniques were addressed in the first section. Chromatographic methods were covered in the second half. Early on in the development of the chromatographic process, it was used to separate compounds depending on their hue such as using herbal pigments and scope of applicability widened over time.

These days, chromatographic techniques are recognized as effective separation technique since they are extremely sensitive, convenient and easy to use. The technique of chromatography permits the separation of a sample mixture into its constituent parts for qualitative (type of material) and quantitative examination (amount of the substance). Using chromatography, a strong technique for isolating individual compounds from a complex sample, a fingerprint profile may be created that can be used to evaluate the sample's whole substance pattern. Several well-established chromatographic techniques for sample evaluation are discussed, including column chromatography, thin-layer chromatography (TLC), high-performance thin-layer chromatography (HP-TLC), microemulsion thin-layer chromatography (ME-TLC), densitometric thin-layer chromatography, two-dimensional thin liquid chromatography, graft thin-layer chromatography and hydrophilic interaction liquid chromatography (HILC).

Conflict of Interest

None to Declare.

Consent for Publication

None.

Acknowledgment

None to Declare.

References

Chen, Z., Turng, L. S., & Wang, K. K. (2006). Adaptive online quality control for injection-molding by monitoring and controlling mold separation. *Polymer Engineering & Science*, *46*(5), 569-580.

Fujii Y., H. Fujii, M. Yamazaki, *J. Chromatogr*. 258 (1983) 147–153.

Harris D.C., Quantitative *Chemical Analysis, sixth ed.*, W. H. Freeman Media, Bedford, New York, USA, 2003.

J. Faghihi, X. Jiang, R. Vierling, S. Goldman, S. Sharfstein, J. Sarver, P. Erhardt, *J. Chromatogr.* A 915 (2001) 61–74.

J. Su, P. Fu, Y. Shen, C. Zhang, M. Liang, R. Liu, H. Li, W. Zhang, *J. Pharm. Biomed.* Anal. 46 (2008) 342–348.

Kim, E. K., Jeong, E. K., Han, S. B., Jung, J. H., & Hong, J. K. (2011). HPLC separation of isoquinoline alkaloids for quality control of Corydalis species. *Bulletin of the Korean Chemical Society*, *32*(10), 3597-3602.

Mander L. H., W. Liu, *Comprehensive Natural Products II: Chemistry and Biology*, Elsevier, Amsterdam, The Netherlands, 2010.

Miller R. J., C. Jolles, H. Rapoport, *Phytochemistry* 12 (1973) 597–603.

Newman D. J., G. M. Cragg, *J. Nat. Prod.* 70 (2007) 461–477.

Orlandini, S., Pinzauti, S., & Furlanetto, S. (2013). Application of quality by design to the development of analytical separation methods. *Analytical and Bioanalytical Chemistry*, *405*(2-3), 443-450.

Snyder L. R., J. J. Kirkland, J. W. Dolan, *Introduction to Modern Liquid Chromatography, third ed.*, Wiley–Interscience, Hoboken, NJ, USA, 2010.

Strobel G., A. Stierle, W. M. Hess, *Plant Sci.* 101 (1994) 115–124.

Tistaert, C., Dejaegher, B., & Vander Heyden, Y. (2011). Chromatographic separation techniques and data handling methods for herbal fingerprints: a review. *Analytica chimica acta*, *690*(2), 148-161.

Wei H., L. Sun, Z. Tai, S. Gao, W. Xu, W. Chen, Anal. Chim. Acta 662 (2010) 97–104.

Xie B., T. Gong, M. Tang, D. Mi, X. Zhang, J. Liu, Z. Zhang, *J. PHarm. Biomed. Anal.* 48 (2008) 1261–1266.

Chapter 3

Physical Methods of Analysis

Malik Saadullah[1,*], Muhammad Asif[2], Jahan Ara Batool[1], Durr-e-Shahwar[3], Mohammad Alqarni[4] and Bashir Ahmad[5]

[1]Department of Pharmaceutical Chemistry, Government College University, Faisalabad, Pakistan
[2]Faculty of Pharmacy, Islamia University of Bahawalpur, Pakistan
[3]Peoples University of Medical and Health Sciences, Nawabshah, Sindh, Pakistan
[4]Department of Pharmaceutical Chemistry, College of Pharmacy, T aif University, Saudi Arabia
[5]Faculty of Pharmacy, Bahauddin Zakariya University, Multan, Pakistan

Abstract

There are analytical procedures that deal primarily with nonchemical methods and techniques useful in establishing the qualitative nature of unknown samples. The choice of analytical techniques for the solution of real analytical problems specifically is governed not only by the characteristics of the individual methods, but also by the alternatives available.

Keywords: physical analysis, separation techniques, solubility, molarity, molality, normality, normal fraction, X-ray diffraction, polarimetry, refractometry and particle size analysis

---

*Corresponding Author's Email: Maliksaadullah@gcuf.edu.pk.

In: Functional Analytical Techniques in Pharmaceutical Chemistry
Editor: Malik Saadullah
ISBN: 979-8-88697-059-3

Introduction

Different methods allow the chemist to enhance the scope of analysis. Therefore, in different cases, accuracy in measurements can be done without demolition of the sample. To examine different complex mixtures quantitatively, insurmountable challenges or difficulties are introduced practically. The analytical techniques discussed in this article are solubility determination, molecular weight determination, viscosity determination, surface tension determination, X-ray diffraction, polarimetry, refractometry and particle size analysis.

Solubility Determination

Solubility may be interpreted as the highest quantity of a substance soluble in a specified quantity of solvent at a specific temperature and is a property of a specified solute-solvent combinations. The National Formulary and U.S. Pharmacopeia explains solubility as the number of ml of solvent needed to dissolve 1 g of solute.

IUPAC defines the term solubility as the scientific composition of saturated solution shown as the ratio of specified solute in specified solvent. Solubility could also be interpreted in units of (concentration) for molality, mole fraction, mole ratio and other units.

Solubility is expressed by the following terms:

Molarity (M): total number of moles of solute in 1 liter of a solution is known as molarity.

Molarity= moles of solute/1 liter of solution

Normality (N): No. of grams (g) equivalent of solute in 1 liter of a solution is known as normality.

Molality (m): The measure of the number of moles of the solute in solution corresponding to 1000 g of solvent.

Molality= moles of solute/kg of solvent

Molar fraction (X): It is defined as the amount of a constituent divided by the total amount of all constituents in a mixture

Basic Concepts of Solubility

The solution may be described as a similar mixture of more than 2 components. In two constituent systems, the constituents in the larger amount is normally called solvent and other as a solute.

Classification of Solutions

On the bases of the saturation, the solutions has three types:

1. *Saturated solution:* Solution in which the highest amount of solute is dissolved.
2. *Super saturated solution:* The solution in which the concentration of solute is high.
3. *Unsaturated solution:* The solution in which the concentration of solute molecules are low.

The solutions are again classified on the basis of parts of solvent required to dissolve the solute (Table 3.1)

Table 3.1. Parts of solvent required to dissolve the solute

Term	Parts of solvent required to dissolve 1 part of solute
Very soluble	<1
Freely soluble	1 to 10
Soluble	10 to 30
Sparingly soluble	30 to 100
Slightly soluble	100 to 1000
Very slightly soluble	1000 to 10,000
Practically insoluble	$\geq$10,000

Depending on the solute nature that is simply dissolved in a solvent, again solutions are classified as

1. *Gases in the liquids:* e.g., Ammonia that dissolves in water.
2. *Liquids in the liquids:* e.g., Alcohol that dissolves in water.
3. *Solids in the liquids:* e.g., KOH that dissolves in water.

Mechanism of Solubility (Dispersibility)

Solute is dissolved in solvent by the three steps, which is quite significant. In the first step, the intermolecular bonds of solute particles breakdown between two adjacent molecules. In the 2nd step, the solvent produces a gap itself to attach the solute molecules. In the third step, the solute molecules fitt in the gap present between the molecules of solvent.

Factors Affecting the Solubility (Dispersibility)

- *Temperature:* Basically, solubility increases with temperature. But in some cases, solubility decreased upon increasing the temperature.
- *Nature of solvent*: Polarity of solute molecules as well as polarity of solvent molecules affects the dispersibility. For example, polar solutes are soluble only in polar-solvent, while non-polar solute is only soluble in non-polar solvent.
- *Effect of pressure*: This factor basically effects the dispersibility of gases. Solubility of gases increased with an increase in the pressure.
- *Polymer's Crystallinity*: For amorphous polymers there is no latent heat change entropy on fusion and when crystalline polymers dissolve, latent change gives energy to melt crystallites, enthalpy requires outweighing enthalpy change produced via the disordering process. Most of the polymers under their melting points are not soluble in all solvents. Solubility parameters are helpful in directing solubility behavior of different polymers.
- *Effect of Concentration:* A problem that the solvent system is acceptable at the highest solute concentration induces insolubility in dilute concentrations like clean-up operations.

Solubility Test Methods

Following are the methods used for solubility testing;

Shake Flask Technique

The most dependable technique of log-P resolution is the shake flask technique. This method was first established fifty years ago and is still seen as a reliable solubility technique. The following steps are involved:

1. The production of sample
2. Equilibration
3. Different separation stages
4. Examination of the saturated solution and remaining solid at the end
5. Data analysis and interpretation.

It is feasible that the balance in shake-flask technique resolves when there is equal clear dispersibility of given samples after different equilibration gaps. Equilibrium can outstretch in 24 hours with an authentic agitation method. It is feasible to observe reported stirring time from 48 hours to 2 weeks. When the efficacious surface area for cessation is enhanced, the procedure can be completed sooner via sonicating samples before equilibrium examination. Amorphous material can be added into samples. This will be useful to produce a super-saturated solution that serves as another solution in problem solving method of slow dissolution rate.

Miniaturized Shake Flask Method

Due to the large no. of compounds involved, it is prudent to estimate the solubility and capability to upgrade output of the methods used. This method was introduced & the results were contrasted against semi-operated potentiometric acid/base titrations in the range of 0.03–30μg/ml. The potentiometric technique is very affordable and is capable to produce a pH solubility profile with 1 single resolution. This method can be helpful for all compounds and large variety of media. Its accuracy & output show superiority to potentiometric methods that are applicable to very poorly soluble compounds. Calculated data showed it to be enough for a 1st estimation of the solubility, but they cannot currently us it as an alternate for experimental estimation at main decision points in development process.

Small Scale Shake-Flask Method

This approach is an advanced genre of the shake flask method presented in the study approached by Bergstrom.et al, where the volume of solvent reduced to 1000, 500, 200, 100 and 50 μl. Non considerable differences were seen in solubility data in the range and volume reduction does not effect accuracy. The study was directed in room temperature (22.5 ± 1°C) and samples added in (Milli-Q) water and were shaken at a speed of 300 rpm. Samples were introverted at different time intervals and structurally different compounds were chosen. The drug samples that dissolved rapidly were introverted after 24, 48 and 72 hours where slowly dissolving drugs are removed after 24, 72 and 144 hours. To decrease the loss of substance and sample volume filtration, steps to restore by ultracentrifugation for 15 minutes and samples were examined by (HPLC).

Equilibrium 96-Well Microtiter Plate Methods

Various studies claimed to examine thermodynamic solubility via applying a solvent evaporation technique. First of all, the compound is liquified in an organic solvent. The next step is by incorporating aqueous medium to the drug for reasons of incubation. Once equilibrium is achieved, separation is attained usually with centrifugation and solubility is found out with HPLC.

Column Elution Method

The method found on elution of test substance against water from micro column is imposed with inert support material, and coated with excess test substance. Water solubility is stated by mass concentration of eluate when it extends a plateau as a function of time.

As names showed, the method employed with columns, involving solute migration through the whole system and solute detection as it came out from the column. This method is comprised of micro-column containing silica gel, glass beads or sand. Specified temperature in the column must present. The eluent is double distilled water. When linked to a recirculating pump the headspace of column has capacity for 5 bed volumes of water and 5 samples. The first five bed-volume are wasted to remove water-soluble wastage.

Pumps can be an anastalsis pump by which a flow of 25ml/hr should be maintained. Micro column can be linked to a leveling vessel with a flow rate 25ml/hr is required. This method is impose on substances with less solubility in contrasted to the shake flask method.

Solubility Enhancement Techniques

The solubility of solute can change by changing the crystal form. The equilibrium of solubility relies on properties or nature of interactions in solid phase and in a solution state. Changes in solid phase can consummate with use of amorphous state of drug or metastable crystal forms, salt formation and co-crystal formation. Solubility of solution can effected by altering the pH of the solution (for ionizable drugs).

- *Solubilization by pH adjustment:* To maintain pH by the inducing buffer solution in sample solution to increase solubility of poorly soluble drugs.
- *Solubilization under particle size reduction:* It reduces the particle size and enhances surface area which enhances solubility.
- *Redundance:* Addition of high amount of 2nd solute to the solution of sample to enhance the solubilization.
- *Solubilization by co-solvents:* The utilization of co-solvents predicted by changing organic compounds' solubility. Co-solvents are partially polar because of hydrogen bond acceptors, givers or both certifying solubility in water. Co-solvents enhance miscibility of non-polar preparations and the reason behind it is that the small hydrocarbon areas of co-solvents decrease the capability of water to remove non-polar solutes. The existence of co-solvents in aqueous solutions results in reduced solvent-solvent attractions, directed to reduce dielectric constant, surface tension and miscibility limitations that reflect solvent polarization.

  Examples: Ethanol, propylene glycol, dimethyl sulfoxide, dimethyl acetamide.
- *Solubilization by surfactants (micellar solubilization):* Surfactants are mainly amphiphilic particles which participate in solution & produce accumulated arrangements called micelles. Two models of

micellization are present. 1st is the two phase model and the 2nd one is the mass achievement model. From a solubilization viewpoint, preparation particles linked with micelles and make soluble preparation molecules. The preparation or drug is persumed to be inserted into the micelle in a two-phase model whereas in mass action models, the drug is supposed to bind with participated surfactant micelles.

Applications

1. Used in the bioavailability of the drugs.
2. Used for common salt purification.
3. Used for development of several dosage forms based on the solubility of the main pharmaceutical ingredients.
4. Manufacturing of baking soda.
5. Used in salting out of the soap.

Molecular Weight Determination

The molecular weight of the compound is resolute by resolution of colligative properties.

These are methods used to determine the molecular weight:

1. Osmometric method
2. Ebullioscopic method
3. Cryoscopic method
4. Vapor pressure lowering technique
5. Mass spectrometric technique

Osmometric Method

This technique is used to find out the mass of molecules that rest on colligative belongings and the number of dissolved molecules is single reason which modify solution. Additionally, osmotic pressure, a raise in boiling point, a decrease in freezing point and a reduction in vapor pressure are depend on colligative assets.

Numerous procedures are used in osmometry:

This method is used to measure osmotic pressure of solution parted by clean solvent via semipermeable layer.

Vapour pressure osmometers are used for determination of osmotically active particles amount which decrease solution's vapor pressure.

Freezing point depression osmometers are precise for determination of solution's osmotic strength, as compounds which are osmotically active decrease the freezing point of solution.

Membrane Osmometers

The main principle in the technique is unrestricted distribution of solute and solvent particles by permeable sheath. The tube, having an extensive, has its opening reduced by a piece of cellophane. Then water is passed through the permeable membrane & produces the osmotic pressure to drive sucrose solution up the tube until equilibrium is produced.

Pfeffer was the first person who explained this method. He made precipitation layer in unglazed porcelain wall cups via copper salts react with potassium ferricyanide. He used that precipitation layer which resulted in separation of sucrose solution inside of cup from water on outside and then he detected that water flowed through outside to inside. It is a chief reflection of osmosis. Pfeffer also defined the *Osmosis* process where particles of solvent passed from semipermeable sheath by less concentrated solution towards more concentrated one. Word "osmosis" invented by Greek which means "impulse." He noticed that the stream was relative to amount of sucrose in cup. Pfeffer noticed that at the time where water was present in cup, the pressure needed to apply in compartment that would remove water from cup and such movement is relative to pressure. If sucrose is present in a closed cup, pressure will be produced in it and is directly related to sucrose concentration. Pfeffer predicted that balance phase where water filtration outside is balanced with inner motion occurs through osmosis. Pfeffer defined *osmotic pressure* as such a pressure which is important to discontinue osmotic flow through barrier which is not permeable by solute. It is generally denoted by π. Osmotic pressure is desired to prevent its flow comparative to pressure that initiated it. Osmotic pressure also shows distinct solitary to semi-permeable membrane which is impermeable to solute however is permeable to water.

Then, the weight of molecule is found out from the given formula:

$\pi = (RT)(C/M)$

Cryoscopic Method

Also known as freezing-point depression. The resolution of weight of the unidentified element depends on the principle of freezing point contrasted to particular solvent contrasts rendering to solute amount and nature which is liquified in it.

Weight of molecule is calculated from the given formula:

$$\Delta F = RT_s^2VC/L_fM$$

This process is initially explained through the Rast method by camphor which acts as a solvent. The apparatus contains a tube made of glass with open sides for the introduction of samples. A glass tube specifically surrounded by a jacket of air and a thermometer is implanted inside it. The glass tube, using the jacket, is implanted in vessel comprising mixture of salt and ice which is cool.

$$M = \frac{k_f X_w X 1000}{\Delta T X W}$$

Here $\boldsymbol{M}$ is molecular weight, $\boldsymbol{k_f}$ is cryoscopic constant, $\boldsymbol{X_w}$ is sample weight, $\boldsymbol{XW}$ is camphor weight, $\boldsymbol{\Delta T}$ is depression in freezing point.

Ebullioscopic Method

The method is called elevation in boiling point. This point may explain where temperature of vapor pressure is equivalent to atmospheric pressure. This apparatus consists of glass vessel with condenser and thermometer.

When using a boiling point method to determine molecular weight, the issue which arises is that we do not know about conformation of solution in interaction with indicator. The solution lacks solvent as vapor and condensates on vessel walls. Concentration is not measured by introduced solvent and solute weight. The technique used for this case to sort out trouble which is supply of energy with persistent speed so the quantity of solvent reduced from solution remains same. Then, the device is adjusted for every solvent with constituents of identified weight. The boiling point reduces when non-volatile solute present in solvent. At high temperature, the boiling point is increased.

$$\Delta T = T - T_0$$
$$\Delta T = k\Delta P$$

where **ΔP** is vapor pressure lowering, then by using following formula molecular weight is determined:

$$\Delta T = RT_s^2VC/L_vM$$

Mass Spectrometry

It is one of the modern physicochemical methods for recognization of compounds. This technique is useful for evaluation of molecules. Recent evolution in ionization and separation of ions made it feasible for implementation of process to variety of mixtures and distinct compounds. Because of its sensitivity and selectivity, MS has huge range of applications. Basic components of mass spectrometer are shown in Figure 3.1.

The mass spectrometry method consist of following principle steps:

1. Sample is turned over to source of ion of mass spectrometry
2. Ions of investigated composites induced and transferred to gas phase
3. Separation of ions and its analysis to their mass to charge ratio
4. Ions are ascertained and signs are documented, kept and transferred to spectrum of mass.
5. Sensors work at little pressure, which needs powerful vacuum-pumping systems.

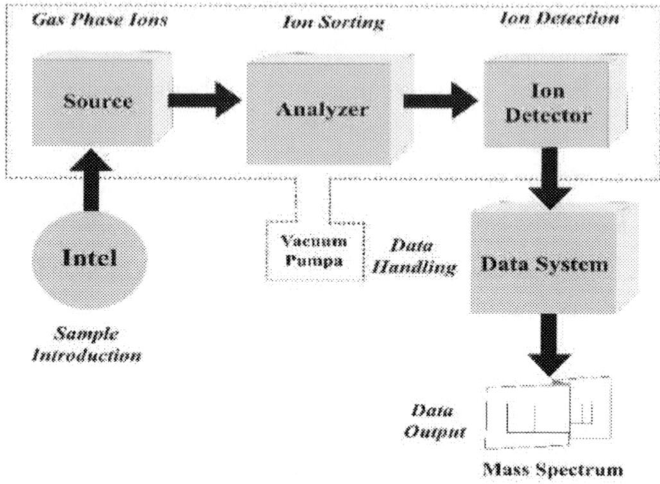

Figure 3.1. Basic components of mass spectrometer.

Viscosity Determination

Viscosity is defined as fluid's friction to flow. Viscosity basically is internal resistance due to molecular forces in fluid that results in friction in flow. It is measured in poise (P), 0.100 $kgm^{-1}s^{-1}$, usually expressed in centipoise (cp), 0.001 $kgm^{-1}s^{-1}$. Viscosity characterizes fluid's friction to flow. The viscosity of water enhances exponentially by decreasing the temperature and is effected by concentration and type of solutes.

Kinematic viscosity explains the absolute viscosity (dynamic viscosity) of substances divided by its density. A high density liquid with less viscosity generally has less kinematic viscosity like fluid that flows quickly through porous phase compared with one with higher kinematic viscosity.

George Gabriel Stokes, in 1851, derived the expression, now known as Stokes law.

$$F_d = -6\pi\mu R V_s$$

where $\boldsymbol{F_d}$ is frictional force, $\boldsymbol{\mu}$ is the dynamic velocity, $\boldsymbol{R}$ is the radius of spherical object.

Factors Affecting Viscosity

Following factors affect the viscosity.

1. *Temperature:* It is directly comparable to viscosity.
2. *Size of molecules:* Size of molecule is directly comparable to viscosity.
3. *Shape of molecules:* Spherical molecules have less viscosity.
4. *Inter-molecular forces:* Larger inter-molecular forces have more viscosity.

Types of Viscometer

Basically, Viscometers range from the classic capillary tube which has no moving parts to advanced hydraulic systems which may measure viscosities at pressures close to 1 GPa (gigapascal) as shown in Table 3.2. Numerous viscometers available in the market like Ostwald Viscometer, Falling Ball Viscometer, Cup-Bob Viscometer, Cone and Plate Viscometer. Typical viscometer is illustrated in Figure 3.2.

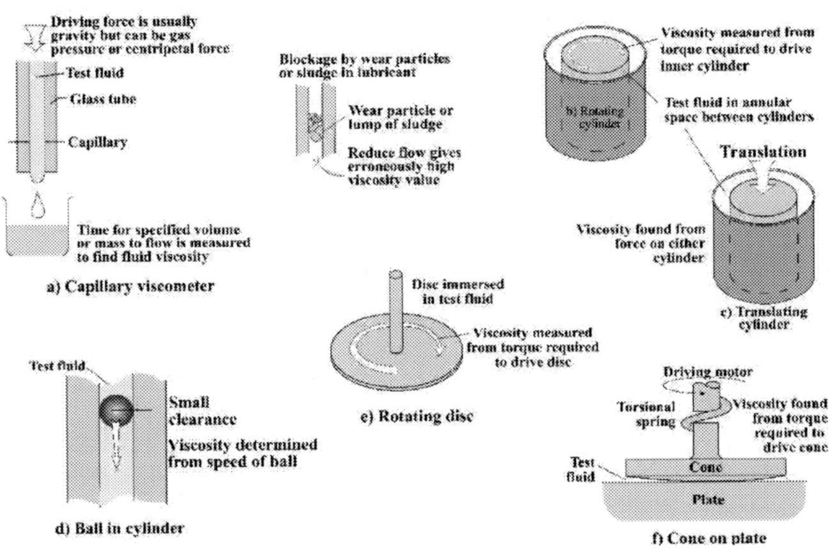

Figure 3.2. Schematic illustration of typical viscometers.

Table 3.2. Viscosity ranges measured by different types of viscometer

Viscometer	η(Pa s^{-1})	Measured unit	Advantages	Drawbacks
Capillary	10^{-4}–10^{1}	Time	Precision	Temperature, $\dot{\gamma}$ control
Falling Ball	10^{-3}–10^{2}	Time	Precision	Temperature, $\dot{\gamma}$ control
Rotational	10^{-3}–10^{6}	Torque	Flexibility, Non-Newtonian fluids	Expensive

Ostwald Viscometer

Basic belief of this technique (Figure 3.3) grounded on flow time for specified liquid volume by precision capillary. It is precise to measure viscosity of liquids at stable or constant temperature. However, by temperature rises, contrasting hydrostatic head does not remain persistent but it would not disturb flow time appreciably by slight temperature interval. Effect is actually important when one has to labor more than extensive temperature variety.

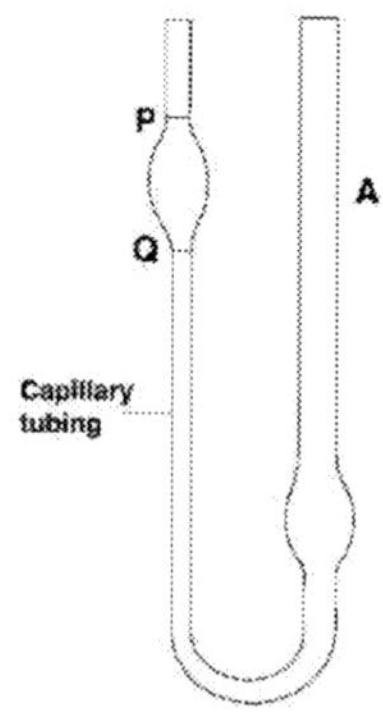

Figure 3.3. Ostwald viscometer.

It may be referred to as a U-tube or capillary viscometer. It is feasible to determine viscosity of fluid with a known density. In a capillary viscometer, water is chosen for the comparison of the viscosity of fluid with standard liquid. A definite volume of water is introduced or added into the limb A of the bent glass tube. This is then forced or sucked into the smaller limb to a height somewhat above the mark P. The water is then allowed to flow back through the capillary tube, and the time t_1 for the level to fall from P to Q is measured. The water is emptied out and the same volume of the liquid to be tested is introduced in the same way into capillary tube and the time t_2 measured for the level to fall from P to Q is then measured again. Preferably the tube should be held vertically in a constant temperature bath with transparent sides.

$$\frac{\eta_2}{\eta_1} = \frac{P_2 t_2}{P_1 t_1}$$

where $\boldsymbol{\eta_1}$ and $\boldsymbol{\eta_2}$ are the coefficients of viscosity of water and liquid respectively, $\boldsymbol{P_1}$ and $\boldsymbol{P_2}$ are the density of water and sample.

This equation is mainly based on Poiseuille's law for a liquid flowing through a capillary tube.

$$\eta = \pi r^4 t \Delta^p / 8lV$$

where $\boldsymbol{r}$ is the radius of capillary, $\boldsymbol{t}$ is the time of flow, $\boldsymbol{\Delta^p}$ is the pressure difference, $\boldsymbol{l}$ is the length of capillary, $\boldsymbol{V}$ is the volume of liquid flowing through the capillary.

Falling Ball Viscometer

It is a commonly and widely used practice in experimentally defining liquid viscosity. It is a modest and low-cost calculating scheme. Falling Ball Viscometer is used in determination of steady shear viscosity of viscoelastic liquids by little shear rates and it entails dimension of velocity of sphere in liquid. It is essential to determine Newtonian solutions. In this technique a sphere-shaped hardened ball is permitted to fallen along tube comprising the sample to be determined.

A spherical shaped harder ball with identified radius $\boldsymbol{r_s}$ and density $\boldsymbol{\rho_s}$ is fallen in fluid. By travelling of sphere descendent by liquid because of gravitational force liquid applies an ascendant drag. Supposing ascendant and descendant services are in equilibrium that is total force from weight of sphere and forces ranges zero and sphere ultimately explore via persistent terminal velocity $\boldsymbol{U_t}$. The drag force $\boldsymbol{F_d}$ employed to sphere falling passed fluid body is explained by "Stoke's law" as:

$$F_d = 6\pi\eta r_s U_t$$

Here $\boldsymbol{\eta}$ is fluid viscosity. Basically, the above equation qualifies usage of this technique. Fluid viscosity may be calculated by given formula:

$$\eta = \frac{2g_s^2(\rho_s - \rho_f)}{9U_t}$$

where $\boldsymbol{\rho_f}$ is density of fluid. Above formula put on simply when radius of sphere is too small as compared to radius and height of tube. It indicates that liquid resistance on wall of tube has insignificant influence on speed of sphere.

Experimental setup of the viscometer is illustrated in Figure 3.4 with implementation of the viscometer apparatus in Figure 3.5.

Two Austria microsystems (Austrian electronic firm which enterprises and assemble sensors for minor form factor, short power, maximum sensitivity and multi-sensor uses) TSL1401CL image sensors used to find out passing the sphere in two sites. This sensor is 128x1 pixel direct monochromatic photodiode rays by consumer contact (frame rate) and sampling period. Productivity for respective pixel is specified analog voltage. LED are positioned in visible portion of respective sensor rays for enlightenment. Space among two sensors is shown with L. Two lined radars connected to NXP MKL25Z128VLK4 microcontroller which offers signals to linear radars and sampled analog outputs of sensors. Algorithm for detection of sample is instigated in microcontroller. Time orders where sphere is identified at topmost and bottommost positions are $\mathbf{t_{Top}}$ and $\mathbf{t_{Bot}}$, correspondingly. Later existences of these two instances, time difference $\mathbf{\Delta t = t_{Bot} - t_{Top}}$ measured and outcome is referred to checking PC through serial communication.

By this system, terminal velocity of sphere may be calculated by

$$U_t = \frac{L}{\Delta t}$$

Flowchart of the implementation of falling sphere measurement is drawn (Figure 3.6).

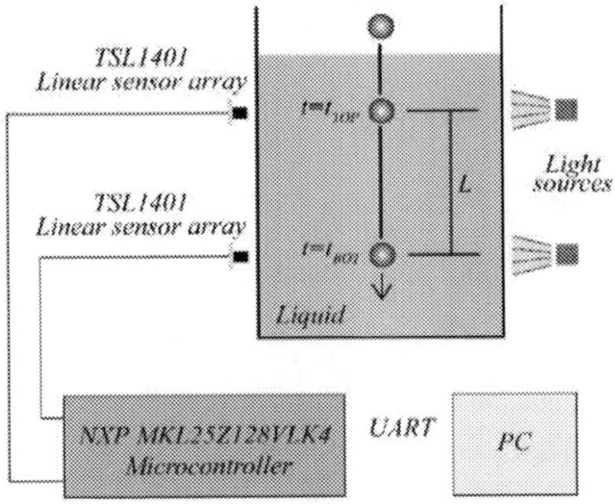

Figure 3.4. Experimental setup of the viscometer.

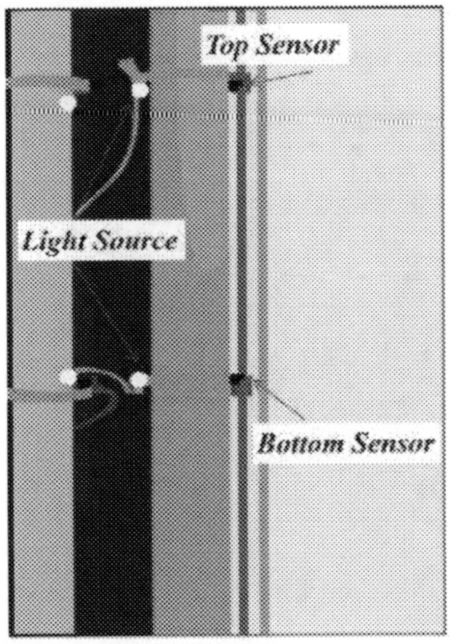

Figure 3.5. Implementation of the viscometer apparatus.

Cup-Bob Viscometer

It denoted to couette viscometer. In couette viscometer the liquid sample whose viscosity needs to be calculated fill up area among two erect coaxial cylinders, the inner cylinder hanges by torsion wire while other one is revolved at persistent speed and then resultant torque on inward cylinder is determined by turning of wire and is a basic principle of viscometer. It is also known as rotational viscometer.

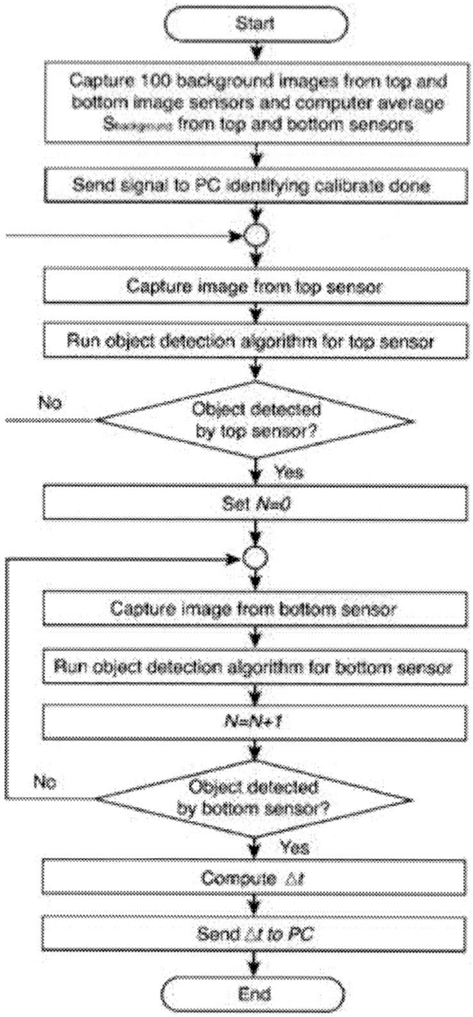

Figure 3.6. Flowchart of the implementation of falling sphere measurement.

Viscosity is calculated from mentioned equation:

$$\eta = K_v T / \Omega$$

here $\boldsymbol{K_v}$ is constant, $\boldsymbol{T}$ is torque, $\boldsymbol{\Omega}$ is angular velocity.

Angular velocity is determined by given equation:

$$\Omega = \left(\frac{1}{\eta}\right)\left[\frac{T}{4\pi h}\right)\left(\frac{1}{R_b^2} - \frac{1}{R_c^2}\right)$$

where $\boldsymbol{T}$ is torque,$\boldsymbol{h}$ is depth of bob immersed in liquid, $\boldsymbol{R_b^2}$ and $\boldsymbol{R_c^2}$ are radii of bob and cup respectively.

Cone and Plate Viscometer

It is used in long range then other instruments of this type because of its simpler construction. With this instrument, liquids with viscosities from 1-50000 poises have examined and rates of shear 0.0002 to $9000\mathrm{sec}^{-1}$ is achieved. A 1000-30000 fold range rates of shear can covered with one filling of viscometer. It was used for studying the rheological characteristics of certain gels. Schematic diagram of the cone and plate viscometer is illustrated below Figure 3.7.

The main advantages of this method measuring system are real. The 1st filling is instant and temperature equilibrium in test sample is obtained in few seconds. Viscometer cleaning is generally time consuming operation. This instrument is cleaned by simply wiping off flat plate and flat surface of cone. The velocity s given by the following formula:

$$\eta = CXT/v$$

where $\boldsymbol{C}$ is the constant, $\boldsymbol{T}$ is the torque, $\boldsymbol{v}$ is the speed of cone.

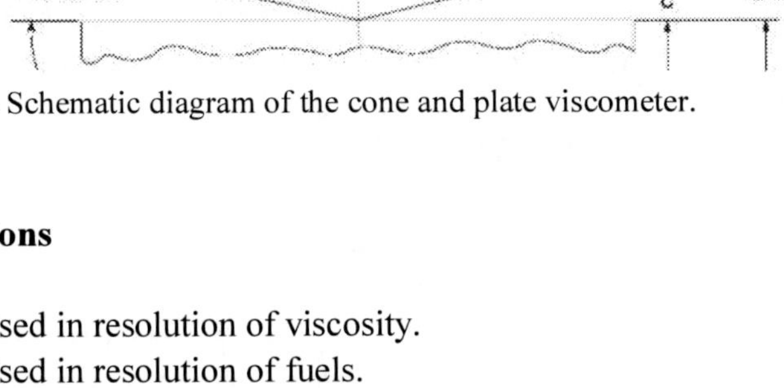

Figure 3.7. Schematic diagram of the cone and plate viscometer.

Applications

1. Used in resolution of viscosity.
2. Used in resolution of fuels.

3. Used in control of polymerization.
4. Used in resolution of flow under different solvents with pumps use in GC or HPLC.
5. Used in study physical properties of a liquid.

Surface Tension Determination

It is interpret as force which is accountable for reducing surface area of liquid, unit of this force is Nm^{-1} or Jm^{-2}. By following equation surface tension (γ) can be calculated as:

$$W = F \times d$$

where ***W***=work, ***F***=force and ***d***=distance.

Therefore,

$$(\gamma) \text{ Gamma} = F/d$$

Status of water droplet over the solid surface affected by two forces, one is gravitational force and the other one is surface tension. The droplet volume is the major factor by which the affecting force is determined. For a drop of water present in a solid surface, interfacial tension forces γ_{SL}, γ_{SV} and γ_{LV} are produced status of the water droplet either flattened over surface or remaining as a sphere, it depends upon balance between these forces.

$$S = \gamma_{SL} + \gamma_{LV} - \gamma_{SV} > 0$$

where S is power of spreading, γ_{SL} is solid liquid surface tension, γ_{LV} is liquid vapor surface tension and γ_{SV} is solid vapor surface tension.

Examples of the surface tension are drop of water, paper clips floating on water and insects walking on water (Jo, Jo et al. 2017). Above 90% of relative humidity (RH), wood and paper can adsorb little amounts of water vapor as free water because of surface tension effects of water.

For many biological as well as industrial processes surface tension of water is important parameter. By studying the formation of water droplets here we show that surface tension of a freshly prepared water surface is even higher (~90 mNm^1) than under equilibrium conditions (~72 mNm^{-1}) with a relaxation process occurring on long-timescale (~1 ms). Dynamic adsorption effects of hydroxides (OH^-)/protons (H^+) may occurred at origin of this dynamic surface tension. However, the dynamic surface tension does not significantly change by changing the pH.

Factors Affecting the Surface Tension

1. Surfactant present in the solution shrinks surface tension.
2. Enhanced temperature declined surface tension.
3. Enlarged molecular speed rises surface tension.
4. More intermolecular potency, greater surface tension.

Methods for Determining Surface Tension

Following methods are used for determining surface tension.

Static Surface Tension Measurement Methods for Liquids

- *Du Nouy ring*
  It is an outdated manner to measure the surface tension and interfacial tension. Surface is used to measure the extreme pulling force on ring. In this method the moistening effects of liquids don not show any impact.
- *Wilhelmy plate*
  The wilhelmy plate performance is majorly relevant towards liquids, after the surface tension is essential to be stated for extensive periods. An erect plate which perimeter is known, binds with balance and is immersed in liquid. That balance is used to take wetting force.
- *Spinning drop*
  This technique is right to find out little interfacial tensions. Thickness of a drop surrounded by a hefty phase is dignified whereas both are being revolved.
- *Pendent drop*
  In this method both forces are taken.

Dynamic Surface Tension Measurement Methods for Liquids

- *Bubble pressure*

Surface tension at small surfaces is measured by this technique. The variable taken in this process is the supreme pressure of every boil.

- *Drop volume*
  The drop volume procedure used for defining interfacial tension by means of time task. When liquid of single density thrusted into subsequent liquid of changed density the time among two following drops is quantified.

Surface Tension Measurement Methods for Solids

- *Sessile drop*
  The sessile drop system is best for rounded models. This procedure chiefly comprises of determining the contact direction optically and used to evaluate the drenching belongings of a restricted area at a solid surface. At end, position among starting point of the droplet and tangent at its edge are resolute.
- *Dynamic Wilhelmy*
  This procedure obliges the even geometry to calculate average advancing and receding contact positions on solids. Equal sides of solid should have equal belongings. On solid material the wetting force is taken as solid is plunged in or quiet by liquid with specified surface tension.
- *Single fiber Wilhelmy*
  It is functioned to sole fibers to quantify progressing and disappearing contact positions.
- *Powder contact angle*
  It is used to determine the usual contact direction and sorption speediness of powders, additional spongy kinds and restrained the modification in weight as a time function.
- *Liquid homolog method (Zisman)*
  The homolog process is created to find out the contact approach of numerous fluids and scheming cosine of approach versus the surface tension of liquids. Critical surface tension of solid is where cosine of the contact angle is equivalent to 1, attained through extrapolation.

Standard Contact Angle Measurement Methods for Various Materials

- *ASTM D724-99*
  It is an accurate assessment process intended for paper surface wettability. The purpose of this process is to determine the

resistance of paper surface wettability by calculating performance of liquid droplet which functioned straight on surface of paper.

- *ASTM D5725-99*
  It is a standard assessment technique normally aimed at surface wettability and sheeted things absorbency by programmed contact angle checker. It determines the contact angle of sample liquid on smooth sample of film.
- *ASTM C813-90*
  This technique is standard for hydrophobic alteration on glass through measurement of contact position. This way, majorly insures the recognition of hydrophobic adulteration on surfaces of glass.
- *ASTM D5946-96*
  It is the standard test technique applied to corona-treated polymer films by measurement of water contact angle. It includes contact angle measurement of drops of water on corona-treated polymer film surfaces. Wetting tension of film is evaluated by results.
- *TAPPI T458*
  It is the test for wettability of paper surface. In it, the direction of contact among air and liquid on surface of paper is reserved to determine resistance on surface of paper to moistening via liquid. Primary angle of contact and early wettability reflected to be measured the paper's ruling worth. The amount of change in wettability reflected a measure of writing quality.

Applications

1. Used to find out the surface tensions of air-liquid or liquid-liquid interfaces.
2. Used to measure altered surfactant critical micelle concentrations.
3. Used to remove dust units.
4. Used in the making of detergents and soap.

X-Ray Diffractometry

X-Ray diffraction (XRD) is a method mostly taken to express organization of crystal atoms in which an X-ray hits crystal and rays of light distribute into several precise angles. Diffraction patterns depend on the speed of X-beams consumed and the assembly of item. One spreads out an atom wavelength radiation to observe an atomic structure. It is restricted on crystals assembly normally exact periodicity and symbolize itself a diffraction design for X-rays created certainly. Though, this technique bounces significant records while studying solids with a reduced amount of ordered structure, like liquids, liquid crystals, amorphous solids, polymers and others.

This method permits fair determination of crystalline materials, compounds like vitamins and antibiotics etc. The thorough structural examination of the crystal often permits to do different chemical chores, for example stating a bond type, chemical formula, molecular weight with identified density, symmetry and configuration of molecules and molecular ions.

XRD is prevailing and the non-destructive performance is used to describe the crystalline tools. It offers broad data about assembly, phases, surface and additional morphological limitations like normal grain size, strain, crystallinity and its flaws.

Principle

In XRD, diffraction of ray from crystals is a reflection of periodicity of crystal manner, thus limitation in the crystal frame frequently effects deprived diffraction things. Mainly, crystal may defined by the help of grid or lattice, distincted with three axis and directions among them. Principle of X-Ray diffraction is schematically presented below Figure3.8.

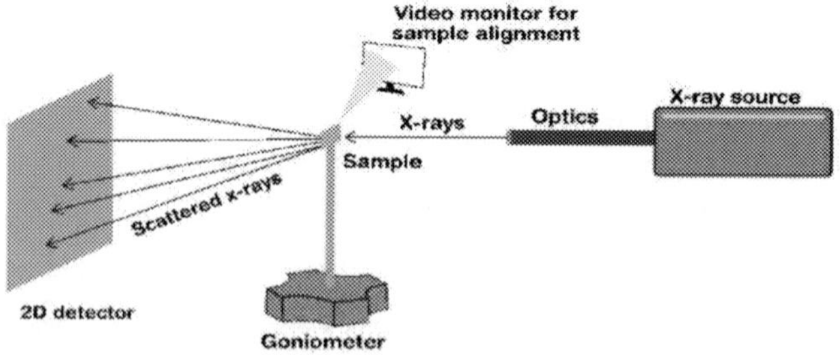

Figure 3.8. Schematic diagram of X-Ray diffraction.

Bragg's Law

$$n\lambda = 2d \sin\Theta$$

where **d** characterize the spacing among diffracting planes, **Θ** is incident angle, **n** is any integer, and **λ** is wavelength of beam. Bragg's Diffraction Law is illustrated below Figure 3.9.

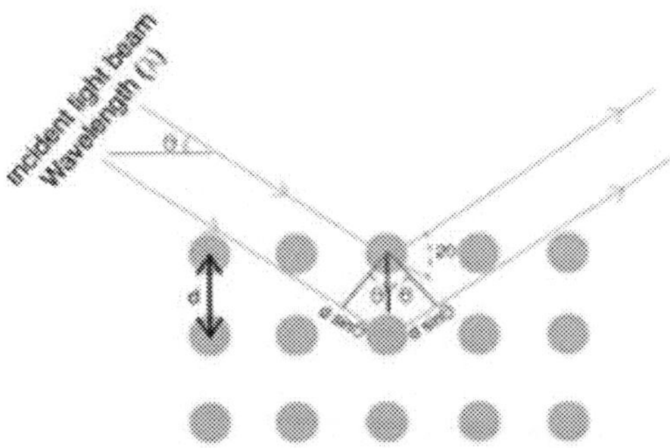

Figure 3.9. Schematic description of Bragg's Diffraction Law.

Thomson Scattering

Scattering of X-ray is normally how to find density of electrons in crystal. Energy of X-rays is more as compared to valence electrons. In scattering like Thomson scattering, the electromagnetic beam relates to extra electron.

X-Ray diffraction is functions magnificently in reviewing polymers' crystalline state. It also provides substantial information while reviewing amorphous solids and liquids. X-Ray outlines the solids comprised of numerous diffused rings, the intensity of which is going on quick fading by zooming. By the shape, width, and intensity of these rings one can determine particularities of the short-range order in a firm liquid or amorphous structure.

Application of X-Ray Diffraction

This technique is one of the main methods which describe crystals structure. Organic and inorganic compounds, polymers, amorphous materials, metals, alloys, minerals, liquids and gases, protein molecules, nucleic acids etc. are studied by this method. Uses of X-ray diffraction (Figure 3.10) are listed below:

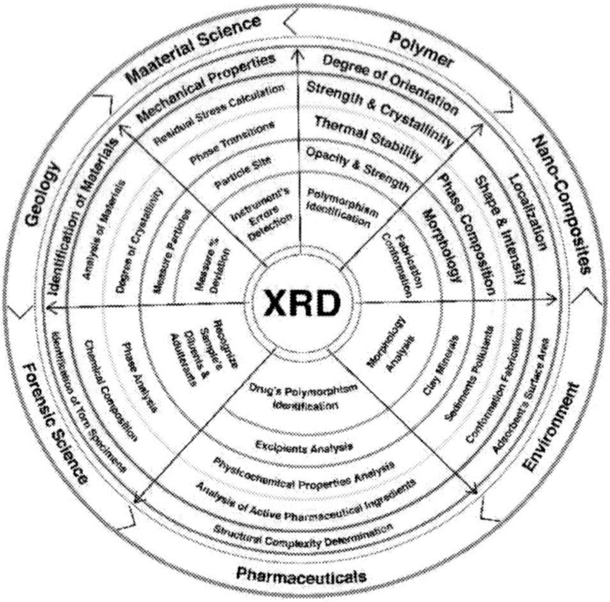

Figure 3.10. Application of X-ray diffraction.

1. Used for the identification of phase.
2. Used in soil survey.
3. Used in petroleum refinery industry.
4. Used to determine particle size.
5. Used to resolve cis-trans isomerism.
6. Used to study corrosion product.

X-Ray Fluorescence Analysis (XFA)

It is the modern spectroscopic way to learn materials in order to attain their essential composition. When constituents are exposed to gamma or X-rays, ionization of particles may occur. Fluorescence is basically a method in which absorption of radiation of definite energy gives re-emission of radiation with altered energy.

XFA used to identify the composition and physical properties of two samples (reference and experimental). This method is significant when there is a need to search individual features of the two samples' composition.

Working

XFA is used to analyse elements that use major rays to blast exterior of tester by radiations (Figure 3.11).

In three-way X-rays interact with matter: Smattering over the item, absorbed through atom and diffusing by material.

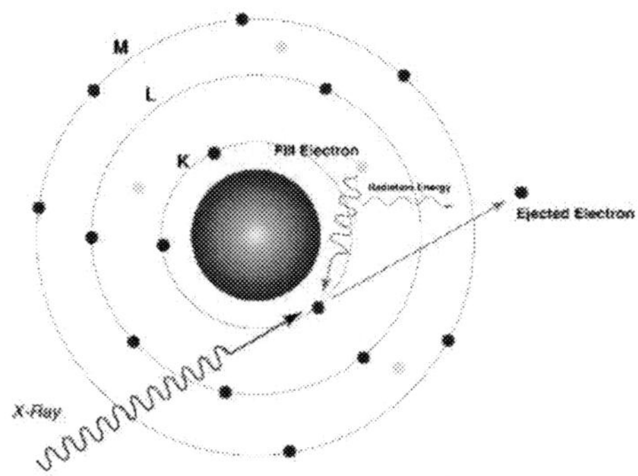

Figure 3.11. Diagram basically shows electron ejected from an atom in the K shell caused by a primary excitation X-ray. The vacancy is filled by an electron from another shell, which emits a secondary fluorescent X-ray.

XFA is material absorption and once an X-ray absorbed, it may result in the ionization of atoms. Ionization of an atom happen when the atom is open to emissions with greater energy than the ionization contains release of one or more electrons. Gamma and X-rays may bounce enough to eliminate strongly held electrons by innermost orbitals. By removal of electron, the structure of atom become unstable and electrons in upper orbitals fall on lower ones to cover the place which is left behind. In falling, energy mainly emits in the shape of a photon. Its energy equals the energy modification of orbitals working in this method. Hence substance releases radiation having characteristic energy of atoms exist.

Applications of X-Ray Fluorescence Analysis

Outlining of heavy metals in soil, qualitative and quantitative investigation of soil, evaluating microelements in soil and agricultural goods, sediments, water, aerosols, minerals, rocks, control of the raw material worth, invention and engineering of out-put method, lead paints examining, determining of oil and fuel contagion, measuring of treasured constituents concentration, shaping toxic metals in foodstuffs, elemental enquiry, study of paintings, archeological discovers age determination Figure 3.12.

Environment	Water	Potable, effluent, river, snow, sea, rainwater, fog condesate
	Air	Dust particles, aerosols, gaseous dust, exhaust gases, light coal ash
	Soil	Deposits
	Biosphere	Leaves, grass, algae, plants, vegetables, fungi, roots, needles
Geology		Microcrystals, minerals, rocks
Medicine	Fluid	Blood, serum, cellular water, sweat
	Tissues	Lung, liver, hair, bones, nails
Industry	Petrochemistry	Oils, lubricants, crude oil, fuel, coal, coal ash
	Metallurgy	Metallic and non-metallic alloys, high-quality aluminium
	Other branches	Thin films of different nature and composition, glass-supported coatings, surficial contaminants of Siwafers, semiconducting materials, ultrapure reagents Radioactive compounds, dust from electric furnaces
Archeology		Mansonry of ancient Rome, ceramics
Forensic examination		Remnants, tissue particles, finger prints

Figure 3.12. Application of X-ray fluorescence analysis in different fields.

Polarimetry

Polarimetry is the ancient and most commonly used procedure constituting a nonseparation performance. It depends on measurement of a net outcome. Data from uncontaminated enantiomers are mandatory for assessment. The reliance of this technique is that the chiral molecules revolve around the plane of polarized light. The optical stuffs of materials are reformed under the impact of external stresses like mechanical stress or stress produced by magnetic or electrical fields. These features can be categorized by polarization techniques. (Klug and Alexander 1974)

Even though polarimetry was the first way to determine enantiomer structure for most of the twentieth century, it is hardly used for any other purpose. Now it is most commonly used to check the absolute alignment of synthetic composites when linked to literature figures. Current computational methods, attached with chiroptical procedures, are being established that allows the calculation of precise rotations and determine the absolute pattern. This method actually measures the revolutions of monochromatic polarized light plane after passing through a sample. A schematic diagram is shown below Figure 3.13.

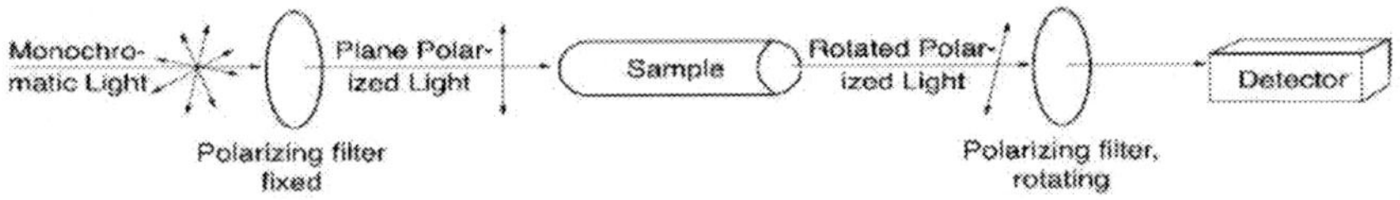

Figure 3.13. Schematic representation of a polarimeter.

Working

The spinning of a plane of polarized light depend on irregular molecules and steric configuration. Spin of a plane of diverged light is directly related to concentration, path length and angle of rotation. It is measured by Biot's formula:

$$\alpha = [\alpha]_D^{25} \times C \times d$$

Here **C** is the concentration of sample solution, **d** is the path length of sample cell, $[\boldsymbol{\alpha}]_{\mathbf{D}}^{\mathbf{25}}$ is the specific rotation of sample at **D** is the line of sodium where temperature is 25^0C.

Features Affecting Optical Activity

1. Nature of sample.
2. Sample concentration.
3. Sample tube length.
4. Wavelength of incident light.
5. Sample solution's temperature.

It is not noticeable why a chiral medium should have this outcome, until the linearly polarized light beam, denoted by the sine wave as shown in Figure a, is cracked into the two spherical polarized constituents shown in figure b and c of Figure 3.14.

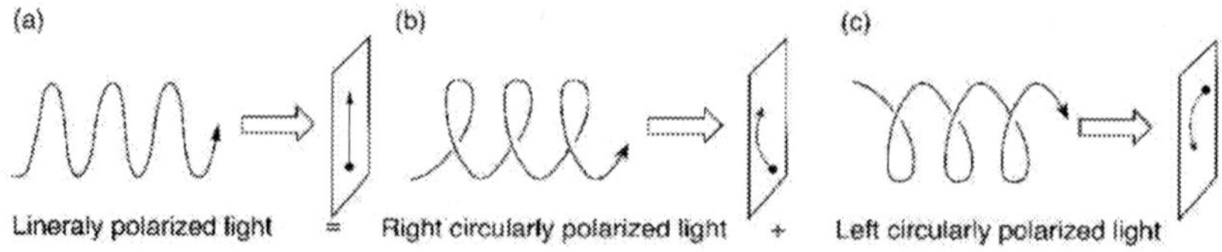

Figure 3.14. Schematic representations of the waveforms of polarized light beams passing through a perpendicular plane.

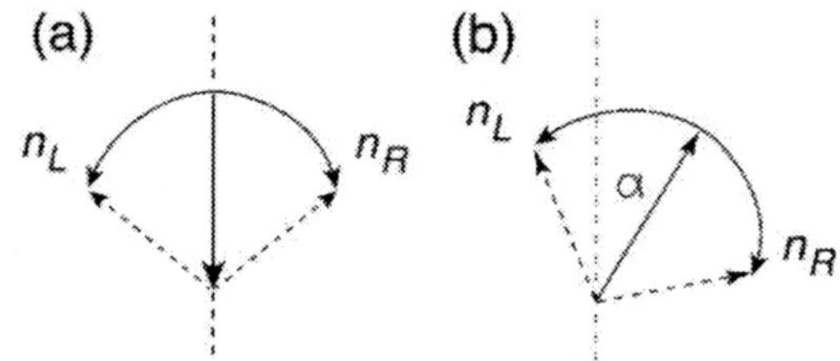

Figure 3.15. Schematic diagram shows the effect on the transmitted plane by refraction of circularly polarized light beams relative to the incident plane of polarized light (dashed line). (a) When $\mathbf{n_L} = \mathbf{n_R}$, the vector sum (solid arrows) of the two circularly polarized beams (dashed arrows) remains in the same plane as the incident beam. (b) When $\mathbf{n_L} \neq \mathbf{n_R}$, the vector sum of the two waves is rotated $\boldsymbol{\alpha}^{\mathbf{0}}$away from the plane of the incident light.

The point of junction moves beside a line, when the linearly polarized beam goes through a vertical plane. The point of intersection defines a circle, moving towards the left or right reliant on the chirality logic of the helix, when the spherical polarized beams pass through the same plane. The vector sum of the right and left spherical polarized beam matches linearly polarized ray. Both left and right polarized beams are refracted likewise by an achiral medium, their refractive indices $\mathbf{n_R}$ and $\mathbf{n_L}$ (which measure change in velocity) are equal. As shown in schematic diagram (a), the vector sum of two deflected circularly polarized rays remain in the plan incident polarized light, that is, the plane is not revolved. But in the case, where $\mathbf{n_L}$ and $\mathbf{n_R}$ are not equal, the two beams are moved out of segment, and their sum of vector is exchanged by an angle $\boldsymbol{\alpha}$, as shown in schematic diagram (b) of Figure 3.15.

Important Tools CD and ORD

In polarimetry, circular dichroism (CD) and optical rotatory dispersion (ORD) are leading and major tools for the learning of features arising from optical activity.

The theory of ORD is like simple polarimetry defined above, normally it measures the optical cycle of complex as a task of wavelength. Circular dichroism (CD) is similar, but it fluctuates in that the substrate holds a chromophore that engages at the wavelengths employed. The molar absorptivities of both right and left circularly polarized rays are diverse, in this exceptional case. As a result, the vectors of the diffused beams are also of uneven magnitude. Since the emergent beam no longer crosses a line, but it defines an ellipse, and the evolving light is elliptically polarized. In the state of CD band, due to the immersion the ORD shows "anomalous" performance (a Cotton effect). The mean wavelength among ORD peak and trough is nearby the λmax of the chromophore gripping the light. It is not uncommon for the ORD arch to change mark in such state. Since ORD processes rotation, it is tentatively finite at all wavelengths, where CD measures absorption difference, it happens only in the area of an absorption band. CD is very delicate and normally used in basic biology for peptide and protein structure enquiry, but seldom in organic production labs. CD can used to regulate two samples having same configuration.

Applications

1. This system used for the quantitative determination in the sugar industry.
2. Used to identify several compounds.
3. Used to determine chiral composites.
4. Used to determine substance's optical transparency.
5. This method mostly used to find out the unknown solutions concentration.
6. Used in proteomics.

Refractometry

Refractive Index (RI) defined the speed of light in a void compared to light speed in a measured medium. RI is chiefly framed as part of the depiction of liquid trials, in the same mode that melting points are habitually obtained to illustrate solid complexes. RI principle elaborates the diversion centered on the speed of light, which authorized through altered mediums, whenever the speed of light deviates as it passes a frontier from one medium to another than its track of travel also deviates. RI depends on the wavelength and temperature of the resolution being dignified.

Refractive index (n) = speed of light in vacuum/speed of light in a substance.

This technique is mainly used for determining the refractive index of liquid which fluctuates regarding humidity content. The water content of liquids is deliberate using a refractometer.

For calculating the amount of moisture in solids, liquids and gases, energy of moisture absorption near-infrared light may be used. At 1.94 μm wavelength, absorption of energy by humidity is elevated, while at 1.7 μm wavelength energy immersion because of humidity is zero. Hence, determining absorption at both wavelengths is allowed due to constituents in solid other than water to be rewarded for and resultant is directly proportional to content of energy. Now, modern instruments use multiple-frequency infrared vigor and have higher proficiency to eradicate the constituents' outcomes in the material and distress the result. Such up-to-date multifrequency apparatuses also handle deviations in particle size in the restrained material in a better way.

Principle

The procedure of the refractometer is constructed on the physical source of light refraction, the index of refraction is calculated by *Snell's Law*. Speed of light reduces as it permits into more optically solid media, and increases as it goes into less optically solid media. The variation in speed is supplemented by a change in track, and at a firm angle of incidence, the light does not divert into the second medium, but is totally or completely reflected. The position at which this transpires, is called the critical angle. This angle is stated by refractometer. A refractometer consists of a light source, cleaned to a single wavelength, which is focused towards the prism-sample boundary by a joining lens. This produces a series of incidence positions, some of which will be replicated totally (those less than the critical angle). By using a charge-coupled device (CCD) sensor, the force of reflected light measured at particular angle α, at which light initiates to be redirected completely is also firmed. RI of the trial can be calculated using the identified refractive index of the prism, since this slant depends on the ratio of the refractive index of prism to that of the sample. Schematic diagram of the revolution of prism or modification of wavelength results are shown in Figure 3.16.

The relation between the refraction of the light between the air and the medium is given by Snell's Law:

$$n = \sin i / \sin r$$

where **n** represents the refractive index, **sin i** is the angle of incidence and **sin r** is the angle of refraction.

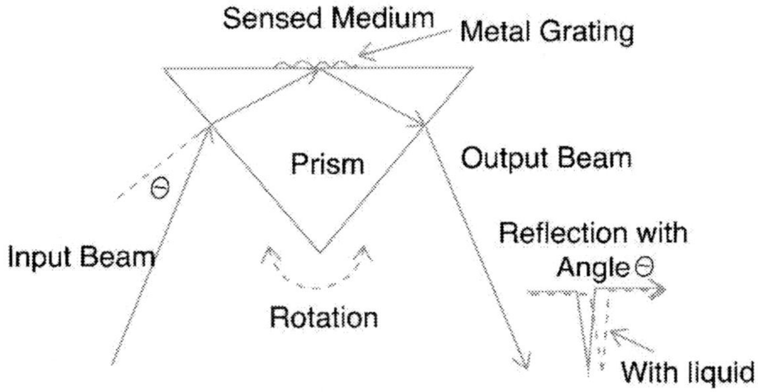

Figure 3.16. Schematic diagram of the rotation of the prism or the tuning of the wavelength results in a coupling to the SPP (Surface Plasmon-Polariton Sensor) at a specific angle. Measurement of the change in the coupling angle (or wavelength) with and without the sensed medium results in the calculation of its refractive index.

Factors Affecting the Refractive Index

- *Temperature:* As the temperature rises, the refractive index declines which means temperature is inversely related to refractive index.

- *Wavelength:* It is a vital parameter that marks refractive index, sodium D-line at 595 nm is the proper wavelength to find the refractive index.
- *Viscosity:* It is also the main factor and inversely related to the refractive index.

1. *Becke line method*
   It was earlier suggested by Mittchell and chiefly helped to determine faceted stones. This procedure contained the microscope with well-lit field brightness. The RI of the indefinite substance is restrained using the known material.
2. *Immersion contrast method*
   It is mainly used for determining the assistance and presence of girdle and facet boundaries once submerged in identified liquid.
3. *Direct measurement method*
   For sole refracting stones this process is effective. It includes the capacity of the RI using the microscope with a vernier balance.
4. *Minimum derivation method*
   This process is the most accurate and contains the use of a table spectrometer for the calculation of the RI. This way needs skill and best settings.

Applications

Measuring whole protein amount in plasma with a refractometer is real, fast, and reasonable method to guess the immunoglobulin amount by inducing it from the entire protein concentration.

1. This technique is used to determine the refractive index of the mixtures.
2. This method is used to determine composites concentration.
3. This method is used to determine structure.
4. This method is used to determine critical micelle concentration of complexes.

Particle Size Analysis

PSA is the physical stuff of compounds which impacts the stability, solubility and process ability. The part for the particle mass is the micrometer. There are various methods to determine the particle size (Figure 3.19).

The range of process is mainly created upon the following:

- Nature of the complex, solubility, toxicity, flowability, cost and specification necessities.
- Types of diameters
- Martin's diameter: basically length of line which sect other line.
- Feret's diameter: Space among two angles of the element.
- Projected area diameter: Thickness of circle consuming identical area as that of the unit.

Methods Used for the Analysis of Particle Size

Microscopy, sieving, sedimentation, sensing zone, light scattering and surface area measurement methods.

Microscopy

Microscopy procedures are unassuming and very beneficial to determine the morphology of units. The main benefits by means of this technique for particle size examination are that it can be constituent specific, and is paired to other particle sizing procedures. This method is time consuming, very low all over and it does not provide the deepness of element. It may be the appropriate technique for particles size determination in some semisolid goods, like ointments and gels. This is supportive in calculating the particle size inside the range of 0.2 μm to 100 μm. The trials via consuming microscopy for particle sizing are (1) determining the number of particles for exact analysis of the invention, and (2) the level of talent essential by the machinist.

Sieving

It was an ancient method used to explore particle size. Chiefly, units are categorized built on magnitude, liberated of any extra unit features like density and surface assets. Sieving is shown diagrammatically in Figure 3.17.

To categorize particles of size series 5–20 μm the micromesh strainers are used while bits with 20–125 μm categorized in average woven wire sieves. Using punched plate sieves the abrasive particles (>125μm) are categorized. Punched plate sieves normally used in industries, sieves may take diverse conformations.

Sieving trials are directed by the help of 11 sieves fixed with gradually higher space openings on way to top and hiring the powder on upper sieve, at the lowermost the receiver is positioned. Numerous patterns for quivering sieves using ultrasonic resources are present. Then, remainders in every sieve noted and stated in percent as accumulative values alongside the nominal sieve gap values.

The chance of particles going from sieve gaps rest on distribution of particles size, height and shape of unit, quantity of particles on the sieve, sieve quaking method and ratio of exposed part of sieve to whole one. Moreover, the procedure of sieving may be exaggerated by friability and cohesiveness of powder.

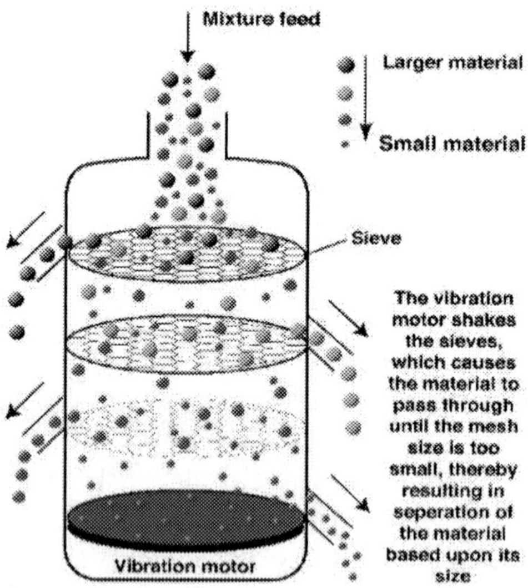

Figure 3.17. Schematic diagram of sieving.

By this method, the powders are categorized into succeeding three types:

- *Coarse powders:* 95% powder clearances from the sieve no. 1400.
- *Moderate fine powders:* 95% powder passes from the sieve no. 300.

- *Fine powders:* 95% powder passes from the sieve no.180.

Several sieve aperture progression ratios commonly available depending on the different international standards (IS), which are quite significant are explained in Table 3.2.

Table 3.2. Table of International sieve standards.

Name of country	Standard	Sieve type
Great Britain	BS 410	Woven wire
USA	ASTM E11	Woven wire
	ASTM E161-607	Micromesh (electroformed)
Germany	DIN 4188	Woven wire
	DIN 4187	Perforated plate
France	AFNOR NFX 11-501	Woven wire
International	ISO R565 1972(E)	Woven wire
		Perforated plate

Sedimentation Methods

This method is generally used for element sizing. It provides complete distribution of particle size examination, because of ordinary sedimentation that is underneath gravity only or enriched the sedimentation rate because of centrifugation. Particle distance, that is stated stoke's diameter, is diameter of identical density range and free-falling hustle by means of element being evaluated.

Basically, measurements of particle size composed at high dilution and caution are essentially supposed to confirm that elements are properly dispersed.

Modest methodology which gains distribution of particle size through cheerfully accessible, low-cost equipment with precision between 2 and 5%, is to custom the Andreasen pipette (Figure 3.18). The Andreasen pipette device contains cylinder-shaped sedimentation bowl of 550 ml with 10 ml fixed pipette by two-way stopper to qualify selection at fixed deepness (20 cm). Well-dispersed suspension lesser by 1% dissolve phase size is hired inside bowl that is kept at persistent temperature using resources of water bath. After these trials of diffusion, the suspensions are withdrawn gently at unvarying time periods (usually in a 2:1 time interval progression) and their items per unit volume are firmed. Designs of the particle size depend on defining early amounts of elements at zero time and amount of sample after time interval T.

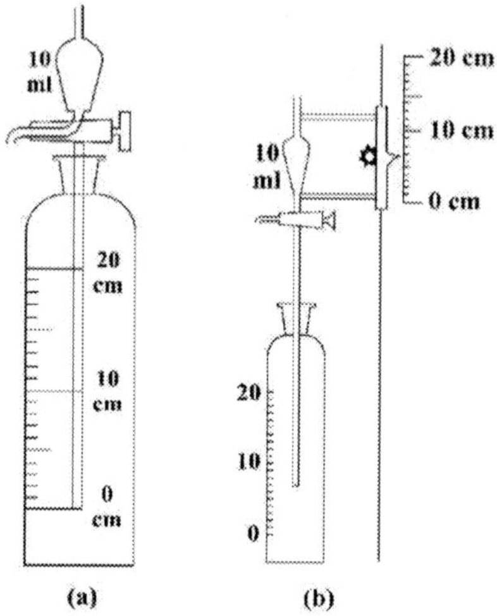

Figure 3.18. Schematic diagram of Andreasen pipette (a) and variable depth modification (b).

Zone Sensing Method

In this method the quantity and extent of the elements which are deferred in the electrolyte solution is firmed by going through the hole which is positioned in the electrolyte solution. When the trial solution passing over the cavity produces a voltage variance which is restrained by the pulse processor. Conductivity counters contain dual chambers. Nanoparticles are present in an electrolyte mixture and positioned in sample compartment and in reference compartment only electrolytic solution id positioned. After this nanoparticle mixture clearances by electrolyte solution. As nanoparticle mixture goes by the electrolyte solution, alteration in conductivity happens. The conductivity variance is the role of a particle size. This process used to determine size of particle in series of 30 to 2000 μm.

Counting means having many profits, comprising of measurement of accurate volume circulation, great resolution and an extensive series of sample dimensions. There are numerous drawbacks of counting approaches, containing need for setting, electrolyte use as medium (limited to hydrophilic particles) and low concentration of particle. Additionally, blunders could arise by porous units, units under the lowest noticeable size unseen and struggle by high-density things.

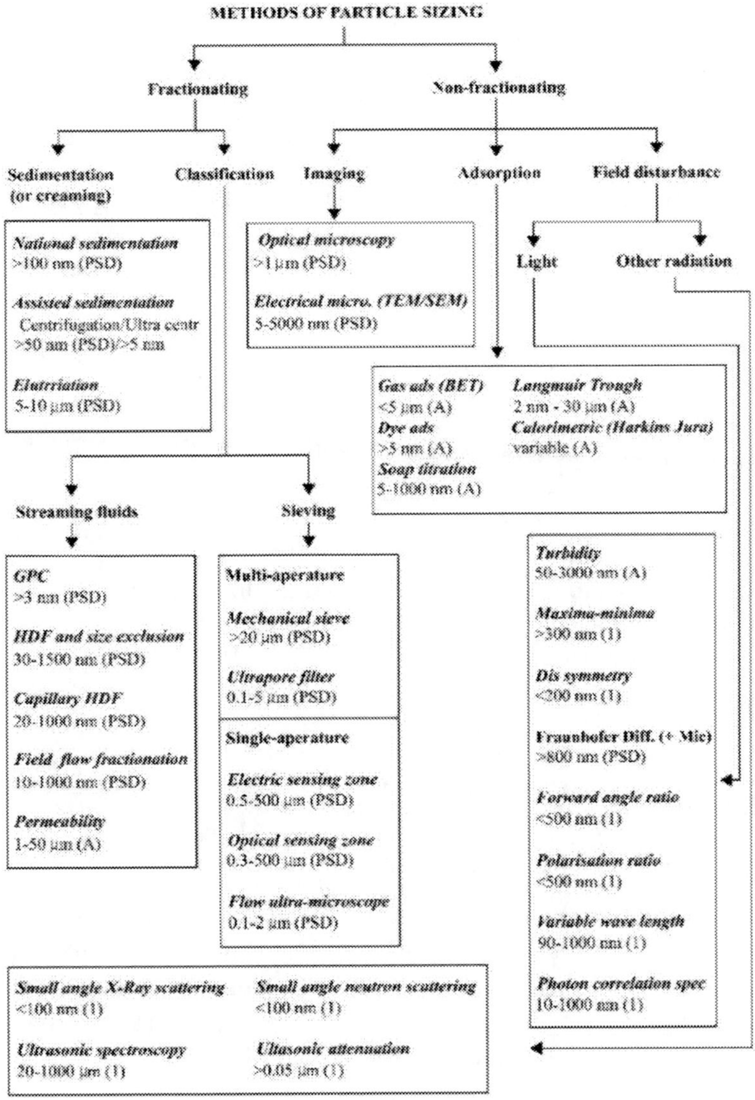

Figure 3.19. Methods used for the analysis of particle size. Where, PSD = particle size distribution obtained; I = average size and indication of size distribution; A = average size only.

Light Scattering Method

Light scattering methods allow noninvasive, quick, and frequently real-time examination of a varied range of ingredients. This technique is mainly created on the aim of scattering. The elements present in the trial may reveal, refract, deflect or engage the incident light. The small elements show a huge angle of scattering and the bulky elements show a slight angle of scattering.

Finding of light smattering from stuff is a beneficial method which claims in frequent precise disciplines which relies on light sources and sensors, definite belongings of molecules can basically be calculated. The main aim of light-scattering trial is that it is visible to monochromatic light wave and a proper indicator senses signal. John Tyndall described single earliest light-scattering experiments, which categorized light smattering by colloidal suspensions (Tyndall effect) in which elements are greater compared to incident light wavelength (Tyndall 1868).

Applications

1. Used in stability lessons.
2. Used to determine the excipient physical limitation.
3. Used for the measurement of flow properties of powder.
4. Used in the determination of product suitability.
5. Used for the determination of processability.

Conclusion

Physical procedures are logical methods that deal mainly with nonchemical devices and practices useful in beginning the qualitative flora of unknowns. This chapter highlights two chief tasks that experts need to organize in order to complete the analysis. The initial task is to complete preliminary procedures that take the system beneath investigation into physical situations suitable for study. The second assignment is to measure physical quantities that can be linked with known arrangements for proof of identity in terms of assembly and society.

Following are acceptable analytical practices: solubility fortitude (Shake-Flask process, Miniaturized shake-flask technique, Small scale shake-flask process, Equilibrium 96-well microtiter plate technique, Column elution method), molecular weight determination (Osmometric method, Ebullioscopic technique, Cryoscopic process, Vapoure pressure lowering technique, Mass spectrometric process), viscosity determination (Ostwald viscometer, Falling ball viscometer, Cup-Bob viscometer, Cone and plate viscometer), surface tension determination (Du nouy ring technique, Wilhelmy plate technique, Spinning drop process, Pendent drop technique, Bubble pressure process, Drop volume technique, Sessile drop method, Dynamic Wilhelmy technique, Single fiber Wilhelmy technique, Powder contact angle system, Liquid homolog method (Zisman), ASTM D724-99 method, ASTM D5725-99 method, ASTM C813-90 method, ASTM D5946-96 method, TAPPI T458 method), X-ray diffraction, X-ray fluorescence analysis (XFA), polarimetry, refractometry (Becke line method, Immersion contrast method, Direct measurement method, Minimum derivation method) and particle size analysis (Microscopy, Sieving, Sedimentation method, Zone sensing method, Light scattering method), as well as their central principles, the necessary equipment, and uses of selected physical procedures have also been debated.

Conflict of Interest

None to Declare.

Consent for Publication

None to Declare.

Acknowledgment

None to Declare.

References

Awang Bono, Krishnaiah, Duduku, Mariani Rajin, Nancy J. Siambun. Variation of Reaction Stages and Mole Composition Effect on Melamine-Urea-Formaldehyde (MUF) *Resin Properties, in Studies in Surface Science and Catalysis,* H. K. Rhee, I. S. Nam, and J. M. Park, Editors. 2006, Elsevier. p. 713-716.

Barton, A. F. M., Solubility parameters. *Chemical Reviews*, 1975. 75(6): p. 731-753.

Bhattacharyya, B. and D. Majumdar, A modified Ostwald viscometer. *Journal of Chemical Education*, 1973. 50(3): p. 194.

Deb, P. K., et al., Chapter 6 - Pharmaceutical and Biomedical Applications of Polymers, in *Basic Fundamentals of Drug Delivery*, R. K. Tekade, Editor. 2019, Academic Press. p. 203-267.

Dembicki, J. H., Chapter 4 - *Interpreting Crude Oil and Natural Gas Data, in Practical Petroleum Geochemistry for Exploration and Production*, J. H. Dembicki, Editor. 2017, Elsevier. p. 135-188.

Edwards, R. A., Chapter 3 - *Viscosity, in Physics for O. N. C. Courses,* R. A. Edwards, Editor. 1970, Pergamon. p. 41-52.

Feher, J., 2.7 - *Osmosis and Osmotic Pressure, in Quantitative Human Physiology* (Second Edition), J. Feher, Editor. 2017, Academic Press: Boston. p. 182-198.

Katyal, A. and R. D. Morrison, Chapter 11 - *Forensic Applications of Contaminant Transport Models in the Subsurface, in Introduction to Environmental Forensics* (Second Edition), B. L. Murphy and R. D. Morrison, Editors. 2007, Academic Press: Burlington. p. 513-575.

Mantri, R. V., R. Sanghvi, and H. Zhu, Chapter 1 - *Solubility of Pharmaceutical Solids, in Developing Solid Oral Dosage Forms* (Second Edition), Y. Qiu, et al., Editors. 2017, Academic Press: Boston. p. 3-22.

Mittal, B., Chapter 2 - *Pharmacokinetics and Preformulation*, in How to Develop Robust Solid Oral Dosage Forms from Conception to Post-Approval, B. Mittal, Editor. 2017, Academic Press. p. 17-37.

Moreno, R., Rheology, in *Encyclopedia of Materials: Science and Technology*, K. H. J. Buschow, et al., Editors. 2001, Elsevier: Oxford. p. 8192-8196.

Pierce, T. B., Physical methods of analysis. *Journal of Radioanalytical Chemistry*, 1972. 12(2): p. 11-21.

Reif-Acherman, S., *The Pre-history of Cryoscopy: what was done before raoult?* Química Nova, 2009. 32: p. 1677-1684.

Stachowiak, G. W., A. W. Batchelor, and G. B. Stachowiak, 9 - *Lubricant or Process Fluid Analysis*, in Tribology Series, G. W. Stachowiak, A. W. Batchelor, and G. B. Stachowiak, Editors. 2004, Elsevier. p. 221-252.

Sudha, P. D. C., *Pharmaceutical Analysis*. 2012: Pearson Education India.

Veseli, A., S. Žakelj, and A. Kristl, A review of methods for solubility determination in biopharmaceutical drug characterization. Drug Development and Industrial Pharmacy, 2019. 45(11): p. 1717-1724.

Viswanathan, P., Y. Muralidaran, and G. Ragavan, Chapter 7 - Challenges in oral drug delivery: a nano-based strategy to overcome, in *Nanostructures for Oral Medicine,* E. Andronescu and A. M. Grumezescu, Editors. 2017, Elsevier. p. 173-201.

Zagorevskii, D., 2.28 - *Mass Spectrometry, in Comprehensive Coordination Chemistry II*, J. A. McCleverty and T. J. Meyer, Editors. 2003, Pergamon: Oxford. p. 367-380.

Chapter 4

Classical Methods of Analysis

Muhammad Asif[1], Malik Saadullah[2,*], Sana Mehmood[2], Munira M Alrooqi[3], Muhammad Hanif[4], Muhammad Uzair[4], Akhtar Rasool[5], Sheikh Abdur Rashid[6], Sania Arif[2] and Bisma Kanwal[2]

[1]Faculty of Pharmacy, The Islamia University of Bahawalpur, Pakistan
[2]Department of Pharmaceutical Chemistry, Government College University, Faisalabad, Pakistan
[3]Department of Chemistry, Faculty of Applied Science, Umm Al-Qura University, Makkah, Saudi Arabia
[4]Faculty of Pharmacy, Bahauddin Zakariya University, Multan, Pakistan
[5]Department of Pharmaceutics, Government College University, Faisalabad, Pakistan
[6]Faculty of Pharmacy, Gomal University, DI Khan, Pakistan

Abstract

For a few hundred years, analytical problems were solved by use of classical titrimetric or gravimetric procedures that reached an impressive development stage in the Western world and Europe, when mineral assets of South Africa were first utilized or mined on a commercial scale. The need for speedy and rapid routine methods, more correct and error-free analysis of intricated materials became more important. The trend towards automated and constant plant control operations, the requirement to improve quality of ores or to attain higher recoveries all resulted in

---

*Corresponding Author's Email: Maliksaadullah@gcuf.edu.pk.

In: Functional Analytical Techniques in Pharmaceutical Chemistry
Editor: Malik Saadullah
ISBN: 979-8-88697-059-3

skilled or professional labor's mounting costs and changes in orders made by economical factors upon the analyst. All of these factors made necessary, modifications to classical methods used by conservationists. These trends are accompanied by various examples.

Keywords: classical analysis, acid, base, pH, gravimetric method, titration, chemical analysis, chemical reaction, colored compound

Introduction

Classical analysis is the wet chemical analysis which constitutes analytical techniques that do not use any instrument (electronically or mechanically) aside from balance. The method is based on chemical responses between a reagent (added to analyte) and an analyzed sample (analyte). Wet techniques depend on product formation of chemical reaction, easily detected and measured from the solution. The precipitated product could be solid or colored.

Acid and Base

Lewis base i.e., electron pair donor and Lewis acid i.e., electron pair acceptor.

$$Al_3^+ + nH_2O \rightarrow [Al(H_2O)_n]_3 + Cu_2^+ + nNH_3 \rightarrow [Cu(NH_3)_n]_2 + BF_3 + NH_3 \rightarrow F_3B - NH_3$$

Acidic oxides or nonmetal oxides

$$SO_3 + H_2O \rightarrow H_2SO_4$$

Basic oxides or metal oxides

$$CaO + H_2O \rightarrow Ca_2^+(aq) + 2OH^-(aq)$$

According to Lewis' description, Lewis bases constitutes centers with high electron density and Lewis acids contain centers of low density electrons. In Lewis base or acid, via electron pair donation they utilize new sigma bond formation to limited electron core in acidic reaction. Lewis bases are identified in almost the same way as Bronsted bases are identified. They frequently include atoms with lone pairs, or nonbonding electrons. Contrastingly, atoms constituting incomplete octet, partial or full positive charges are usual in Lewis acids. Lewis acids or base examples are carbocations and water respectively.

pH

Reactions take place in nature, aqueous media's alkalinity or acidity, depend (stated by Arrhenius definition of acids and bases) on concentrations of ($\mathbf{H_3O^+}$) hydronium and (OH) hydroxyl ions.

Acid ionization or dissociation equation:

$$HA(aq) + H_2O(l) \rightarrow A^-(aq) + H_3O^+(aq)$$

Base ionization or dissociation equation:

$$B(aq) + H_2O(l) \rightarrow BH^+(aq) + OH^-(aq)$$

Hydroxide acid ion (base acid conjugation)

Hypothetically, pH is defined as a decimal logarithmic reciprocal of (aHþ) ionic activity of hydrogen in that medium:

pH ¼ log 10 aHþ ¼ log 10 1 aHþ

pH in solutions (very diluted) with proton concentration is directly related:

pH ¼ log 10 cHþ

Buffer Solutions

An aqueous solution containing weak acid mixture with a conjugated base, or contrariwise, is considered as pH buffer solution. The ability to retain constant pH by adding a small strong base or acid amount is a distinctive feature of buffer solution. One reason to utilize buffer solutions in different chemical procedures is to avoid pH solution changes. In a small pH range where living organisms grow using buffer solution i.e., blood to maintain their constant pH. This exceptional buffering solution attributed to equilibrium presence between HA acid with conjugate base A i.e., HA⇄Hþ

pH Measurement Methods

There are following methods for measurement of pH.

Application of Indicator for pH Measuring

It involves two procedures; Firstly, standard color comparison by using solution (buffer) with pH (known) or indicator color immersed in test liquid. Another procedure comprises pH test paper preparation rinsed in indicator, and later immersed into liquid (test). After that, compare color in parallel with standard one. The procedure is quite easy, yet makes an error and a high degree of accuracy degree is not expected.

Flaws

1. Error due to elevated salt content in liquid (test).
2. Fallacy because of test liquid's temperature.
3. Fallacy due to organic matter in test liquid.
4. The pH of high-purity water is not determined by the indicator method unless indicator influence is itself much larger.

Method for Hydrogen-Electrode

In platinum plate or wire, platinum black is introduced to make a hydrogen electrode. In a test solution, the hydrogen electrode (Figure 4.1) is submerged and electric current initiates to solution and then solution becomes saturated by application of hydrogen gas. In electrodes of platinum black or silver chloride, electrode potential is assessed, and is inversely related to solution pH. To assess solution's pH, hydrogen-electrode method considered as standard among different methods. Derived values by utilizing other procedures becomes definitive or dependable if similar to those measured by the hydrogen electrode method. However, because of the time and money required, as well as the inconveniently handling hydrogen gas or marked effect in test solution of compounds that are strongly reduced or oxidized, this procedure is not suitable for everyday usage.

Quinhydron-Electrode Method

Quinhydrone in solution discrete into quinone or hydroquinone. Since quinone solubility varies with solution pH, pH is determined by measuring the voltage difference between reference or platinum electrode. Despite its simplicity, the procedure was not performed accurately in the presence of reduced or oxidized substances due to its rare use, or in the case when test solution's pH is above 8.

Method of Antimony-Electrode

In this technique, the reference electrode and polished antimony rod tip is submerged in test solution, and pH is measured from the potential difference between them. Since, the method is widely used because its apparatus is durable or handled easily. Still, it has low reproducibility and results differ depend on polish degree of electrode, so this method is of limited use.

Method of Glass-Electrode

The glass electrode method operates two electrodes to measure the pH of a solution by voltage (potential) determination between them i.e., reference or glass electrode.

This method is utilized for pH measurement and contributes less impact on results with oxidized or reduced substances, as its potential quickly attains equilibrium or expresses good reproducibility, and method is applicable on various solution types. Not only in industry, but also in many other fields, the glass electrode process is commonly used.

JIS in its "Methods of pH Measurement" section recommends use of glass electrode for industrial pH measurement as hydrogen electrode is not considered appropriate for this purpose.

Semiconductor Sensor Methods

A glass electrode is replaced by a semiconductor chip in the semiconductor pH sensor or ion sensitive field effect transistor i.e., ISFET, which has been in development since 1970 and is easily miniaturized or unaffected by resistance. Miniaturization allows the user to measure smaller sample amounts and to perform measurements in solid state surfaces or very small spaces. This sensor is widely applicable in measuring medicine and in biology fields.

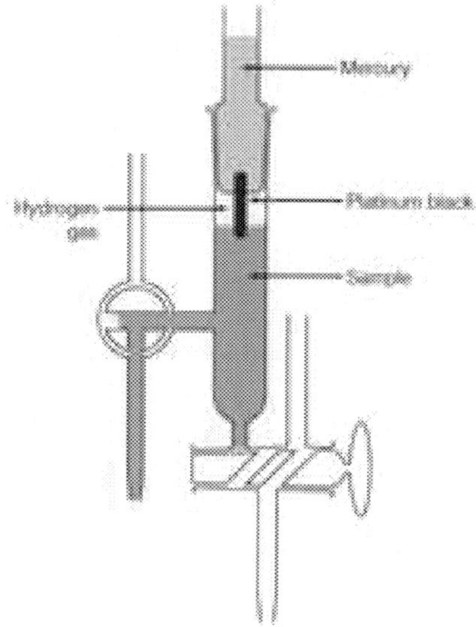

Figure 4.1. Hydrogen-Electrode.

Oxidation Reduction Reactions

Electrons are transmitted in two species (oxidation-reduction i.e., redox reaction), in chemical reaction form. Oxidation-reduction reaction is a chemical reaction in which the ion, molecule or atom's oxidation number varies by the loss or gain of electrons.

$$Fe(s) + O_2(g) \rightarrow Fe_2O_3(g) + Fe(s) + O_2(g) \rightarrow Fe_2O_3$$
$$Ag(s) + H_2S \rightarrow Ag_2S(g) + H_2(g)$$

Titration

A reagent's volume reaction with analyte is measured with this technique. The reagent solution increment i.e., titrant with accurate known concentration, is added to the analyte solution until completion of the reaction occurs between analyte and reagent. Titrant is introduced either from burette (manually) or from an automated titrating apparatus (automatically), so the titrant volume is managed during titration at all times. Titrimetric reaction is illustrated as:

$$x[\text{analyte}] + y[\text{titrant}] \rightarrow \text{product}$$

x and y are the number of moles of analyte and titrant required for titration completion. Titration is finished when analyte reactants are transferred to products and titration has attained an equivalence point. The reagent's consumed amount is read on burette, and the analyte quantity in the sample solution is calculated by x or y values.

Nonaqueous Acid-Base Titrations

When titrated acid or base is infirmed, acid or base titrations, media efficient in aqueous environments, means that acidic pKa should not be above 6.0. Likewise for bases, pKa must not be above 8.0 when utilizing a color indicator to notice an endpoint. By noticing the potentiometric endpoint, basic and acidic pKa values down to 6.0 and up to 8.0 respectively, titrate in an aqueous environment. In case infirmed acidic or basic titration occurs in water (act as a weak base or acid), the aqueous environment must compete for titrant. According to following reaction, the protonation of water is:

$$H_2O + H^+ \rightarrow H_3O^+$$

competing in weak base titration:

$$RNH_2 + H^+ \rightarrow RH^{+3}$$

Deprotonation of water by reaction:

$$H_2O + B \rightarrow OH^- + BH^+$$

competing in weak acid titration:

$$RCOOH + B \rightarrow RCOO^- + BH^+$$

Weak acidic or basic titration happens in solvents rather than water or in solvents by enhancing character i.e., acidic or basic. For weak bases, the main group connection with drugs usually implement anhydrous acetic acid titration rather than water. Weak bases are proton acceptors (strong) and anhydrous acetic acid is considered a proton acceptor (very weak). Hence, anhydrous acetic acid never competes with weak proton bases. According to equilibrium, the protonation of anhydrous acetic acid is due to very strong acids:

$$CH_3COOH + HA \rightarrow CH_3COOH^{2+} + A^{+}$$

Perchloric acid in anhydrous acetic acid is strongest among common acids, and is used to titrate weak bases or perchloric acid to be disintegrate in anhydrous acetic acid. Titration of weakly bases (B) is as follows:

$$B + CH_3COOH^{2+} \rightarrow BH^{+} + CH_3COOH$$

Chelatometry or Complexometric Titration

The volumetric analysis method evaluates titration's endpoint by using colored complex. Titration determines the unknown analyte concentration commonly employed in laboratories which is then identified and is also applicable in quantitative chemical analysis.

Volumetric analysis like volume measurements play a pivotal part. For a reagent used as a standard or titrant, that titrant volume has been reacted. Complexometric titration techniques are applied in solutions for mixture detection having different metal ions.

Reaction quickly attains equilibrium upon addition of every titrant drop and the chance for interfering situations is not possible. Complexometric titration identifies much more accurately the equivalent point. EDTA is widely used as a titrant and has a prolonged usage history.

Calmagite or Eriochrome BlackT are indicator examples that transform from blue to pink by interacting with calcium and magnesium. The complexometric EDTA titration endpoint uses Calmagite or EBT as an indicator which detects change as colour switches from pink into blue.

Argentometric Titrations

Precipitation titration techniques based on reactions yield low solubility compounds (ionic). Silver nitrate is commonly used as a significant precipitating reagent, and is used to determine halogens, like fatty acids, divalent inorganic anions, mercaptans and anions. Silver nitrate titrations are known as argentometric titrations.

The following methods are used in argentometric titrations:

Mohr Method, Chromate Ion

In the Mohr method, (Na_2CrO_4) sodium chromate is utilized as an argentometric titration bromide indicator, as well as for chloride and cyanide ions. In the equivalence-point region, precipitate formation of brick-red color silver chromate (Ag_2CrO_4) is done by reacting silver ions with chromate. The Mohr method should be used at pH of 7-10. At low pH, the concentration of the chromate ion in the acidic solutions is so low that it cannot form a precipitate with Ag+ at the equivalence point. Because in acidic solutions CrO_4^{-2} is converted to $Cr_2O_7^{-2}$ (Chromate ion is conjugated base of chromic acid).

At high pH, silver ions will precipitate as AgO.

1. Titration Reaction:

$$Ag^+ + Cl^- \leftrightarrow AgCl(s)$$

2. White at the equivalence point:

$$2Ag^+ + CrO_4^{-2} \leftrightarrow Ag_2CrO_4(s)$$

(Indicator Reaction) brick-red.

Volhard Method, Iron (III) Ion

Fe (III) ion is used as an indicator in this method. Indirect halide ions determination is common application of Volhard method.

To begin, a predetermined standard amount of silver nitrate solution is introduced into sample.

$$Ag^+ + Cl^- \leftrightarrow AgCl\ (s)\ \text{white}$$

Back-titration along with standard solution of thiocyanate determines excess silver:

$$Ag^{+} + SCN^{-} \leftrightarrow AgSCN\ (s)$$

Excess $AgNO_3$ Iron (III) works as an indicator in back-titration reactions. Due to the excess of thiocyanate ions, the solution exhibits a red color by forming $Fe(SCN)^{2+}$:

$$Fe(SCN)^{2+}.\ Fe^{3+} + SCN^{-} \leftrightarrow Fe(SCN)^{2+}$$

The strongly acidic environment of the Volhard titration for halide ion titration has a significant benefit rather than halide ion titrations due to carbonate, oxalate, or arsenate ions. The ionic silver salts are soluble in acidic media and slightly soluble in neutral media. Silver chloride has more solubility than silver thiocyanate. Hence, by using the Volhard method to determine chloride, the significant reaction results extend to reach the end of back-titration.

$$AgCl(s) + SCN^{-} \leftrightarrow AgSCN(s) + Cl$$

Reaction creates a fade end point, resulting in the occurrence of thiocyanate ion overconsumption.

The low chloride results are resolved by silver chloride filtering before performing back-titration. Other halides do not need filtration as they produce silver salts (having lower solubility than silver thiocyanate).

Fajans Method, Adsorption Indicator

Fajans Method utilizes organic compounds i.e., adsorption indicator adsorbs or desorbs onto the solid surface, in precipitated titration. Desorption along with adsorption occurs ideally near the equivalence point resulting in not only color change but also color transfer from solution to solid or occurs vice versa. Fluorescein is a popular adsorption indicator used in chloride ion titrations with silver nitrate. Fluorescein is partially ionized in aqueous solution to hydronium ions and yellowish green coloured negatively charged fluoresceinate ions, this fluoresceinate ion forms a dark red silver salt. Before equivalence point, the excess Cl^{-} layer is primarily adsorbed. This repels anion indicator, and the secondary adsorbed ionic (counter) layer loosely retained as cations (Na^{+}). Far off equivalence point Ag^{+} is in excess, so the precipitate's surface is charged positively due to the formation of Ag^+ primary layer. In the counterlayer, it attracts and adsorbs an indicator. Titrations require reliable, precise, and fast adsorption indicators, but they're only useful for the few precipitation titrations that form colloidal precipitates rapidly.

Redox Titrations

Redox, acid or base titrations are deployed on proton transfers considered as redox reactions or, if utilized in titration, referred to as redox titration. The following equation is used to represent redox titration:

$$\text{Ox1} + \text{Red2} \longrightarrow \text{Red1} + \text{Ox2}$$

The subscript 1 denotes the material to be titrated, and the subscript 2 denotes titrant. Substance reduction potential is a measure of its ability to absorb electrons. Greater reduction potential of positive value shows:

The substance reduced easily or is a powerful oxidizer by simply removing electrons having reduced the reduction potential from other substances.

A greater substance with reduction potential oxidizes the compound having reduced reduction potential. Generally, titration does not require high equilibrium constant to explain the completion reaction and 100% product formation. In redox titrations, the difference (DE) in two substances or reduction potential of equilibrium constant is determined.

Titration Involving Potassium Permanganate ($KMnO_4$)

Reduction of ($KMnO_4$) i.e., potassium permanganate in sulfuric acid by hydrogen peroxide occurs and it suitably measures H_2O_2 aqueous solutions ranging from 0.25-70 wt.%. Substance reduces $KMnO_4$ and produces positive interference under acidic conditions.

Safety Precautions

Follow safe and handling guidelines, specifically related to personal protective clothing or footwear. Aprons composed of PVC or neoprene are approved, as are neoprene rubber gloves or mono-goggles. Sampling bottles containing H_2O_2 must not be stoppered or sealed, but rather with aluminum foil and paraffin film loosely covered or vented. Solutions removed from a sampling bottle should not again be returned to the bottle, but should be flushed down the drain with enough water. Combustible materials or spills that come in contact with H_2O_2 must flushed right away with water. Adsorbent media must not be used for cleaning H_2O_2 spills.

Reagents

1. Concentrated sulfuric acid involving reagent grade H_2SO_4 i.e., (sp.gr. 1.84).
2. Standardization of potassium permanganate (0.3 N) against sodium oxalate (NIST).

Procedure

Transfer an adequate H_2O_2 aliquot sample to a tared bottle then weigh it on an analytical balance, mentioned in table below. Carefully wash the weighed sample with distilled water into a 250 mL volumetric flask, dilute up to the mark with rigorous mixing. In a 400 mL beaker, pipette 25ml aliquot carrying 250 mL decontaminated or filtered water with 10 mL concentrated H_2SO_4 (sulfuric acid). To obtain a persistent pink color, titrate it by adding $KMnO_4$ (0.3 N) that acts as a titrant in the process of redox titration. Because of giving a bright purple color, $KMnO_4$ works as its own indicator. Keep in mind that the endpoint is obtained when the solution remains a bit purple.

Iodine Redox Titrations

A pharmaceutical product's shelf-life introduction must be tested. Hydrogen peroxide is steadily reduced to water or photodegrades in disinfectants. Vitamin C is inherently unstable in vitamin tablets, or steadily oxidized because of air. The concentration determination of active ingredients is definitive or error-free for many formulations which is most significant. Hydrogen peroxide content in pharmaceutical disinfectants or ascorbic acid concentrations in vitamin C tablets (supplement) is determined in lab. Both analyses utilize an automatic titrator by involving aqueous iodine potentiometric titration with sodium thiosulfate. The electrode of a platinum ring indicator follows the titration curve progress by potentiometry.

Titrations Require Iodine

Iodine; a moderately weak oxidizer is reduced to form the iodide anion, given as:

$$I_2(aq) + 2e^- = 2I^- (aq) \quad E° = 0.621 \text{ V}$$

Since redox reaction is entirely reversible, the iodide anion known to be a relatively weak reducing agent; produce iodine by reacting with oxidizer analytes. Iodine titrations developed for various oxidizing or reducing agent analysis. Reducing agents produce iodide in analysis of iodimetric titration during iodine reaction:

Iodimetry

$$A_{ox} + I_2 A \longrightarrow A_{red} + 2I^-$$

$\mathbf{A_{ox}}$ or $\mathbf{A_{red}}$ is oxidized or reduced analyte forms, respectively. Analyte (oxidizer), in iodometric titrations, reacts with measureless iodide excess to make iodine:

Iodometry

$$A_{red} + 2I\text{-}A_{ox} \longrightarrow I2 \text{ excess}$$

In reaction, the amount of iodine formed is proportional to the original analyte amount of sample. The iodine may then be titrated to determine the concentration of analyte in the sample.

Thiosulfate; a universal iodine titrant; react quantitatively as:

$$2S_2O_3^{2-} + I_2 \longrightarrow S_4O_6^{2-} + 2I-$$

So, iodine reactions utilize for relatively strong oxidizing (through reaction with excess iodide in iodometry) or relatively strong reducing agents (in iodimetry, by reacting with I_2) analysis.

Ascorbic Acid Analysis by Iodimetry

Ascorbic acid; by food nutritionists and pharmacists, known to be anti-oxidant, vitamin C or reducing agent. Iodine quickly oxidizes ascorbic acid i.e., vitamin C, $C_6H_8O_6$, by producing wdehydroascorbic acid, or $C_6H_6O_6$:

$$C_6H_8O_6 + I_2 \longrightarrow C_6H_6O_6 + 2I^- + 2H^+$$

Vitamin C; water-soluble compound, or undeviate iodimetric titration is regarded as standard method for ascorbic acid analysis in various citrus fruits or vitamin tablets. In our experiment, actually iodimetric back-titration is performed in which excess measured iodine is generated in sample solution then with sodium thiosulfate, unreacted iodine is titrated. This back-titration considered as iodimetric titration, based on analyte reaction with aqueous iodine; but, continuous usage of this titration permits same titrant usage i.e., sodium thiosulfate as used in hydrogen peroxide iodometric analysis.

Hydrogen Peroxide Analysis by Iodometry

H_2O_2 or hydrogen peroxide solution are broadly sold as disinfectants. The hydrogen peroxide acts as an oxidizer in redox reaction, reacts with iodide:

$$H_2O_2 + 2H^+ + 2I^- \longrightarrow I_2 + 2H_2O$$

The reaction is not sudden or indirect, however fast enough to form a practical analysis base by iodometric titration. In acidic solutions, constitute immeasurable excess iodine amount, sample is introduced. After reaction completion, iodine titration is done with sodium thiosulfate produced by above reaction.

Titration of Ceric Sulfate

Principle

Determination by titrating H_2O_2 content is done with standard ceric sulfate that changes to pale blue at the endpoint along with usage of ferroin indicator.

Scope of Application

This method shows suitability to characterize H_2O_2 content (1-30%) in aqueous (i.e., Caro's or peracetic acid) or equilibrium mixtures. By diluting the sample, higher concentrations can be measured.

Reagents

1. Ceric Sulfate ($H_4Ce(SO_4)_4$) of 0.1N. Standardized.
2. Ferroin indicator. First weigh $FeSO_4$. $7H_2O$ (0.174 gm) and add 25ml distilled water into it. Stir it to dissolve $FeSO_4$. $7H_2O$, then 0.37 gms of o-pHenanthroline monohydrate is added with continuous stirring to prepare a uniform solution.
3. Sulfuric acid (1:19). Cautiously 50ml sulfuric acid (concentrated) introduced in distilled water (950ml). Mixture of reaction is stirred constantly.

Procedure

Measure sulfuric acid (150ml) in conical flask (500ml), then during first titration, introduce enough crushed ice to retain temperature below 100°C. Add ferroin indicator (2-3 drops) then titrate with ceric sulfate (0.1 N), colour changes to pale blue. In a glass weighing bottle, measure a suitable mass of sample to within 0.1 mg. Let the mass taken be W gms. Then add and mix weighed sample to conical flask's cold contents. Titrate to achieve blue end-point with ceric sulfate. Required titrant amount considered as A mLs.

Gravimetric Methods

To measure analyte, all gravimetric procedures rely on some final weight determination. Gravimetric analysis is a potentially precise class of provided analytical procedures because weight measurement with good precision is suitable rather than any fundamental property. These procedures are considered among the oldest of analytical methods, and are tedious or time-consuming. In order to eliminate interfering substances, samples can need intensive treatment. There are 3 basic gravimetric analysis types: analysis of precipitative or thermo-gravimetry, and electrodeposition.

Precipitation Gravimetry

An insoluble compound develops in solution by introducing precipitant or precipitating reagent to analyte in technique of precipitation gravimetry. Precipitation, simply a metathesis reaction between precipitant and analyte in many cases; however, reaction work as gravimetric method by causing precipitate potentially. Thermal analysis technique or TGA obtained by the sample's temperature increment over time by plotting weight averse to temperature, hence measuring material transformation effectually where at contrasting temperatures, compounds form, evaporate or degrade. Thermal Gravimetric Analysis is employed to determine product's breakdown or development, failed testing or analysis to specify materials like; rubbers, petrochemicals, mono or polymers, compounds i.e., inorganic or organic, and research:

1. Degradation of temperature.
2. Absorbed moisture content and volatiles in materials.
3. Organic and inorganic components level in materials.
4. Chemicals and materials decomposition points or stages, as well as explosives.
5. Temperatures for which solvent remnants are driven off.
6. Loss on Ignition LOI and Loss on Drying LOD measurements on small sample masses like gravimetric techniques.

Thermogravimetric Analyzer (TGA); an essential or important laboratory tool for material characterization. TGA is a powerful tool utilize for material characterization, also employed in distinct polymers, composite materials i.e., CFRP or GFRP, pharmaceuticals, petrochemical or pharmaceutical applications, considered a powerful method by combining with other materials or chemicals, techniques like; Headspace SEM/EDX or GC-MS on coatings (plastics and paint), pyrolysis GC-MS, FTIR for example.

Thermal Gravimetric Analysis implementations include:

1. Quality Control QA / Quality Assurance QA must ensuring products to encounter material specifications.
2. Using TGA to ensure product safety.
3. Determine compound or material's carbon content.
4. Shielded operating temperature recognition of various gases.
5. Product formulating process enhancement.

6. Determination of Pre-exponential factor, activation energy of Arrhenius or kinetic parameters of material is determined when they obey kinetics of first order.

Electrogravimetry

In electrogravimetry, under controlled conditions, the analyte is electroplated or can stay deposited on an electrode, the direct measure of the amount of analyte in sample is the increase in mass.

This technique has the advantages that, the co-deposition of metal can be avoided, if the conditions are carefully controlled. The filtration is not required for this process. When current is passed in electrochemical solution, the most positive reduction potential will occur at the cathode first. For example, if the solution contains Cu (copper), H (hydrogen), and Cd (cadmium) ions, the Cu ions will deposit first on cathode. Because the electrode potential becomes more negative, until the potential is reached when H ions are reduced and H_2 gas is formed at the cathode. The potential at the cathode will remain constant until H_2 gas is generated, which is usually generated until all of the water has been electrolyzed, so that the cathode's potential cannot become sufficiently negative to cause deposition of cadmium ions. The metal ions with lower electrode potential and with higher electrode potential can be separated by using electrodeposition technique. The electrolytic separation of Cu and Ni from cupronickel can be achieved by controlling pH and electrolytic conditions. Copper is determined in strongly acidic solution at a potential not exceeding 4 V (generally 2–3 V), as nickel may also plate out above this potential. For complete deposition, the solution should contain a mixture of HNO_3 and H_2SO_4. However, concentrated acid is not recommended because it may lead to incomplete deposition of Cu or the deposit may not adhere satisfactorily to the cathode.

The respective electrode, following reactions occur:

At the anode,

$$4OH^- \rightarrow 2H_2O + O^{2+} + 4e^-$$

At the cathode,

$$Cu2^+ + 2e^- \rightarrow Cu2H^+ + 2e^- \rightarrow H_2$$

When all the copper is electroplated, then, dilute the remaining solution with concentrated ammonia (iron if present is precipitated as its hydrated oxide and removed by filtration) and on second cathode, Ni (nickel) is plated at 4 A and 3-4 V. For measuring the deposition completeness, continue the electrolysis by adding water, or test it with dimethylglyoxime.

Miscellaneous Types of Analysis

The assay performed for primary aromatic amines is nitrite titration. The apparatus of electrometric nitrite titration consists of open titration vessel with two platinum electrodes, with a potential difference of 50-100 mV, connected by a suitable circuit. A sensitive current measuring device should be included, with 0.1-1 nA sensitivity, usually with an indicating needle in the circuit. A suitable mechanical or magnetic stirring device should be provided in the titration vessel, or a stream of nitrogen passing through the solution may be used to mix the solution. The platinum electrodes with 0.5 mm diameter and 20 mm length are suitable for this circuit. Before each use the electrodes should be cleaned, by immersing them for a few seconds in boiling HNO_3 (~1000 g/L) TS, and 1 mg/mL of ferric chloride R which has previously been added, and then carefully washing them with water.

Recommended Procedure

Mix 50 ml of water and 20 ml of HCl (250 g/l) in a titration flask, then add a small quantity of test substance, and a catalyst, if it is indicated in monograph, then stir well until the substances are dissolved. After that cools the solution at 15°C and titrate with sodium nitrite (0.1 mol/L), perform this titration slowly. The precaution for this procedure includes, continuous stirring of solution during the addition of titrant, temperature should be maintained at 15°C. When the titration is within 1 mL of the estimated end-point, than add titrant in 0.1 ml portions, allowing not less than 1 minute to elapse before adding subsequent portions. At first, the pointer of the measuring device shows deflection after each addition of reagent, then returns back to its original position. No deflection is observed when the end-point of the determination is reached.

Fats and Fatty Oils Analysis

The lipids, fats, and oils comprise of a huge number of organic compounds including fatty acids, monoacylglycerols, diacylglycerols, triacylglycerols (TGs), eicosanoids, phospholipids (PLs), resolvins, sterols, docosanoids, carotenoids, sterol esters, vitamin A, vitamin E, fatty alcohols, wax esters and hydrocarbons. In classical definition, lipids are defined as those compounds which are soluble in organic solvents. This is a loose defination that may include a variety of non-lipid organic compounds. In 2005, a novel definition and comprehensive system of classification of lipids were proposed. The novel definition is chemically based and defines lipids as small hydrophobic molecules that can be produced wholly or in part by thioester and/or isoprene unit's condensation. This novel definition enables lipids and their properties, catalogued in that way which is compatible with other macromolecular databases.

Acid Value (AV)

The Acid Value is the number of mg (milligrams) of potassium hydroxide (KOH) required for the neutralization of fatty acids in 1 g of sample.

$$RCOOH + KOH \rightarrow RCOOK + H_2O$$

The suds appears faster but with less stability, when the material has higher AV.

The suds appears slower but with more stability, when the material has lower AV.

A lower AV means more cleansing (detergency).

AV is only used on fatty acids to estimate SV. For fatty acids, the AV is very close to the SV. AV normally runs about 2 points lower than the SV.

Procedure

In 50 ml of flask, add an accurately weighed 10 g of test substance, equivalent amount of alcohol and ether (neutralized to phenolphthalein by 0.1 N NaOH (sodium hydroxide)), and dissolve the test substance. If in cold solvent, and the test substance does not dissolve, then connect the flask with a condenser and warm slowly with constant shaking. Perform this until the test substance dissolves completely. After that add 1 mL of phenolphthalein TS, and titrate with 0.1 N NaOH VS until the solution remains faintly pink after shaking for 30 seconds. Determine the acid value or the volume of 0.1 N alkali, required to neutralize 10.0 g of free fatty acids (test substance), whichever is appropriate. The more dilute titrant is used if the volume of 0.1N NaOH VS required for the titration is less than 2 ml, or may be the sample size adjusted accordingly. The results can be conveyed in terms of the equivalent volume of 0.1 N NaOH or the volume of titrant used. Before titration, gently reflux the alcohol-ether solution for 10 minutes if the oil has been saturated with carbon dioxide for preservation purpose. Before weighing the test substance, place the oils in a shallow dish under a vacuum desiccator for about 24 h if they were saturated with CO_2 (carbon dioxide) for preservation purpose.

Saponification

Saponification is a "hydration reaction in which the ester bond between fatty acids and glycerol of triglycerides is broken down by the free OH (hydroxide)," both of which are soluble in aqueous solutions. Particularly, this process chemically degrades the lipids which are not soluble in aqueous solutions.

Its uses include the cleaning of lipids which are present in process areas involving cell processing and cell growth, such as the cell disruption process, Saponification Value (SV) and fermentation of bacteria.

The SV is the number of KOH (potassium hydroxide), taken in mg mandatory for the neutralization of free acids and to saponify the esters present in 1.0 g of the substance.

Procedure

In a 250 ml flask, add 1.5 g to 2 g of substance, accurately weighed, and alcoholic KOH, about 25 ml. Continually rotate and heat the flask on steam bath, attached with a suitable condenser for the maintenance of reflux, for 30 minutes. Then add 1 ml of phenolphthalein TS to the solution and titrate the solution (containing excess KOH) with 0.5 n HCl. Then perform a blank test under the same conditions. The titration can also be carried out potentiometrically. The value of saponification is calculated as explained in the above formula, as the difference in volumes, in ml, of 0.5 n HCl used by the actual test and used by the blank test, and multiply the answer by 56.1 and with exact normality of the 0.5 n HCl VS, and divided with the mass of substance, taken in g. Before weighing the test substance, place the oils in a shallow dish under a vacuum desiccator for about 24 h, if they were saturated with CO_2 (carbon dioxide) for preservation purposes.

Iodine Value (IV)

The IV represents the number of grams of iodine absorbed by 100 g of substance in recommended conditions. Unless otherwise specified in the individual monograph, determine the iodine value by following methods:

Method I (Hanus Method)

Procedure

To 250 ml of iodine in a flask, add 10 ml of chloroform (CCl_4), 25 ml of iodobromide TS, and accurately weigh the sample. Secure the flask with a stopper, place it away from light and shake the solution occasionally for 30 minutes. Then put 100 ml of water and 30 ml KI (potassium iodide) IS in specified order. The iodine is then liberated, titrate the remaining solution with 0.1 N sodium thiosulfate VS. After each addition of sodium thiosulfate shake the flask thoroughly. When the colour of the iodine changes to yellow, put 3 ml of starch TS, and continue the 0.1 n sodium thiosulfate VS titration until the blue colour vanishes.

The second test, which is the blank test, is performed in the same way and with the same reagents. The formula which is given below is used for the calculation of iodine value:

$$[126.9(Vg- Vs)N] / 10W$$

In the above formula, the 0.1 n sodium thiosulfate VS volumes used by the blank test and the actual test, in ml, are denoted as ***Vg*** and ***Vs*** and **126.9** is the atomic mass of iodine; and the weight of substance, in g, is denoted as ***W***, and ***N*** is the exact normality of the sodium thiosulfate VS. [NOTE- If more than half of iodobromide TS is absorbed by the substance then repeat the determination with a smaller portion of the substance].

Method II

KI Solution

Dissolve about 10 g of KI in the water to make 100 ml and store it in light-resistant containers. Starch indicator solution: To make a thin paste mix, add 1 g of soluble starch with sufficient cold water, then add 100 m of boiling water with stirring. Mix and then cool it. Only the clear solution should be used.

Procedure

If the sample is not already molten or liquid, melt it. [NOTE-The increase in temperature should not exceed more than 10 from melting point of sample] For the removal of impurities and moisture pass the sample through filter paper, of about two pieces. The filtration may be performed in an oven that contains air and temperature sets at 100°C and complete this task within 5 minutes to 30 seconds. In this process, use clean glassware and a completely dry sample. After completion of the filtration process, place the sample for sometimes until it reaches 68 to 71°C temperature before weighing the sample. When sample has achieved the 68 to 71°C temperature, in 500-ml iodine flask, add the accurately weighed sample by using weighing accuracy associated in table. [NOTE-In the weighting process of substance, the iodochloride TS of 50% to 60% should be added in excess in substance, that is, 100% to 150% of the amount absorbed] Put (1:1) of a fresh mixture of glacial acetic acid and cyclohexane, of about 15 ml, and spin, so that sample become dissolved. Then add 25 mL of iodochloride TS to the flask and secure it with a stopper and spin it to mix. Then wait until temperature reaches at 25°C, protected from light, shake at constant interval of time for 1.0 or 2.0 hours, the time interval depends on the iodine value (IV) of the sample: shake for 1 h for an iodine value less than 150, and shake for 2 h for an iodine value equal to or greater than 150. Then, add 150 mL of recently boiled and cooled water and 20 mL of KI in above solution, then mix it well. All of this should be done within 3 minutes. Then the liberated iodine is titrated with 0.1 N sodium thiosulfate VS, for 30 minutes, and stirred continuously after each addition. 1 to 2 mL of starch solution is added when the yellow colour of iodine has almost vanished, and continue this titration until the blue color is discharged. The blank experiment is performed with same reagents and with same manner at the same time. At the end, the iodine value is calculated, as the volume's difference of 0.1 N sodium thiosulfate used by blank test and used by actual test, multiply the value by 1.269, and divide the answer by weight, in g, of the sample.

Hydroxyl Value

The hydroxyl value is defined as; the hydroxyl (OH) content of 1.0 g of substance is equal to the number of mg of potassium hydroxide (KOH).

Pyridine-acetic anhydride reagent is produced by taking 1 volume of freshly opened or distilled acetic anhydride and 3 volumes of freshly opened or distilled pyridine, mix them well and prepare this just before the use.

Procedure

The accurately weighed quantity of the substance is transferred to the 250 mL conical flask, which has a glass stopper, then add pyridine acetic anhydride reagent of about 5 mL into the flask. For the preparation of blank reagent, take another conical flask of 250 mL with stopper and put pyridine acetic anhydride reagent of about 5 mL into the flask. Then on reflux condenser put both flasks in a steam bath heat for 1h. After 1h, add 10 mL of water and heat them for 10 minutes more on the same steam bath. Cool the solution and add about 25 ml butyl alcohol and neutralized phenolphthalein by 0.5 N alcoholic KOH of about 15 ml to each condenser, and the sides of flasks are washed by remaining 10 ml portion, and remove the condenser. After that, to each flask add 1 mL phenolphthalein TS, and titrated against 0.5 N alcoholic KOH, and quantify the volume expended by the residual acid in test solution as T and volume expended by the blank solution as B. After that, in a 125 ml conical flask put 10g of substance, 10 mL of freshly distilled pyridine and 1 mL of previously neutralized phenolphthalein. Mix them well and titrate against 0.5 N alcoholic KOH. Record the volume which is consumed by the free acid in the test sample, denoted it as A, than compute the OH value by using following formula:

$$(56.11N/\ w)B + (WA/C) - T$$

In the above formula, the **W** and **C,** in g, are weights of substances taken for acetylation and for free acid determination. The exact normality of alcoholic KOH is denoted as **N** and **56.11** denotes the molecular weight of KOH.

Ester Value (EV)

It is defined as the amount of KOH, in mg, required for the saponification of esters in 1.0 g of the substance. If the SV and AV have been calculated, the EV is the difference between these two values.

Procedure

In a 250 ml flask, add 20 to 30 ml of neutralized alcohol and 1.5 g to 2 g of substance, then mix this solution. After mixing, put 1 mL of phenolphthalein TS and titrate against 0.5 N alcoholic KOH VS, perform this procedure until the neutralization of free acid is completed. After that, put 0.5 N alcoholic KOH VS of about 25 ml to flask and proceed as guided under SV, heat the flask, and stop the further addition of phenolphthalein IS. Then calculate the ester value as explained above in definition, the volume's difference of 0.5 N HCl utilized by the actual test and by the blank test and multiply the value with 28.05 and divide the answer by the weight of sample which is taken in grams.

Acetyle Value

It is defined as the "amount of KOH, in mg, required to neutralize the acetic acid (obtained by saponification of 1g of acetylated fat or wax)".

Analysis of Alkaloids

Alkaloids were discovered about 4000 years ago and are extensively used for different therapeutic purposes. Their division is based on their biosynthetic precursor and the heterocyclic ring system, such as tropane, quinolone, isoquinoline, purine, indole, and imidazole among others. They have numerous pharmaceutical properties, so they are the main subjects of various studies. In 1819, German chemist Carl F. W. Meissner first introduced the term "alkaloid", this is originated from the Arabic word al-qali, which is the name of plant from which first time soda extraction was achieved. They are the compounds of low-molecular-weight and about 20% of secondary metabolites of plants are comprised of alkaloids. The number of alkaloids which are derived from different genera of plant kingdom is about 12,000.

Classification Established upon the Biogenesis

Alkaloids exhibit a wide range of variation, in botanical, and biochemical origin, in structure and in pharmacological action. There are a variety of classification schemes that can be used in this regard. The alkaloids are divided into types on the basis of their molecular precursors, their structures, and origins, and also on the biological pathways by which they are obtained.

Alkaloids are divided into three categories:

1. Protoalkaloids
2. Pseudoalkaloids
3. True alkaloids

True alkaloids are derived from amino acids but pseudoalkaloids are not directly derived from amino acids.

Protoalkaloids

The nitrogen of amino acid is present in the form of an alkaloid but it is not the part of hetrocyclic ring system of alkaloids. Their major precursors are L-tryptophan and L-tyrosine and their minor group is structurally made up of simple alkaloids. The major alkaloids in this type are yohimbine, mescaline, and hordenine. They are used for the treatment of various disorders, like mental illness, different pains, and neuralgia.

Pseudoalkaloids

They are known as pseudo, because they are not derived directly from amino acids, but have connection with amino acids as amination and transamination reaction from precursors of amino acid. They can also be produced from nonamino-acid precursors. The phenylalanine or acetate are the sources of derivation of these alkaloids. Common examples are ephedrine, caffeine, and Capsaicin.

True Alkaloids

They are amino acid derived alkaloids and share a heterocyclic ring (N-containing). They have a high level of pharmacological activity and are extremely reactive in nature. They form salts which are water-soluble, and are mostly crystalline in nature, forming a salt as they conjugate with acid. But the nicotine is an exception, it is present as brown liquid. Almost all of them have a bitter taste and are solid in nature.

Classification Established upon the Ring Structure

Tropane Alkaloid

The tropane (C4N skeleton) nucleus is present in this group of alkaloids. The Solanaceae family contains a large number of them. The ornithine and acetoacetate are the main source of their derivation. Pyrolines are the structural precursors of these alkaloids. Many of them are esters of mono, di, trihydroxytropane and have wide range of hydroxylation activities. Most common derivatives are cocaine, atropine, and scopolamine, and these derivatives have various pharmacological activities, so have been widely studied since 19th century.

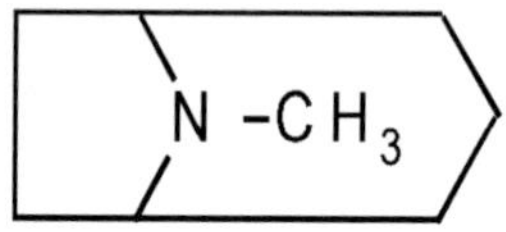

Pyrrolizidine Alkaloids

This group of alkaloids is distinguished by its pyrrolizidine nucleus. They are occurred in the Asteraceae and Fabaceae families of plants. Many of these alkaloids occurs in plants as N-oxide, but during isolation and purification, this role becomes lost. They are toxic alkaloids especially for the liver. In the food chain they act as antifeedants for animals. The most well-known alkaloid of this type is Senecionine.

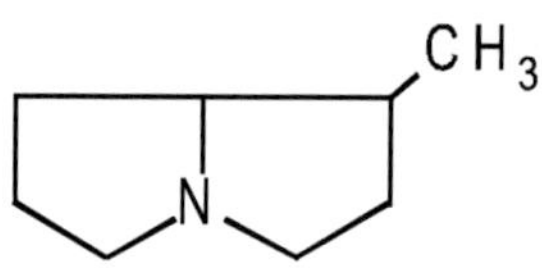

Piperidine Alkaloids

The ring system of these alkaloids is piperidine ring nucleus. The piperidine is a lysine-derived alkaloid. The monocyclic compounds which contain nucleus of C5N are the key feature of these alkaloids. Piperidine alkaloids are characterized by the presence of an odour. These alkaloids exert chronic neurotoxicity. With respect to origin, many of them originated from plants, some from acetate, some from acetoacetate and some from simple pyrrolidine.

Quinolines Alkaloid

The Cinchona plant's bark is the only source of this kind of quinolone-nucleus-containing alkaloid. However, basic heteroaromatic quinolines are obtained from several marine sources, like, form cephalopod ink the 4.8-quinolinediol is obtained and from marine pseudomonad the 2-heptyl-4- hydroxyquinoline is obtained. Cinchonine, cinchonidine, quinine, and quinidine are the main alkaloids in this group.

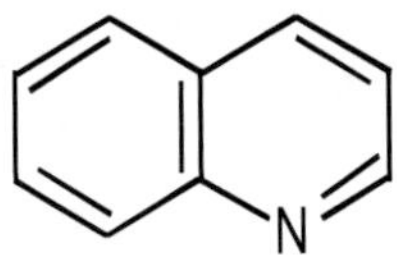

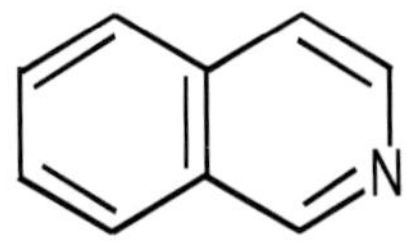

Isoquinoline Alkaloids

These alkaloids comprise the largest group of alkaloids and are found in higher plants and some are present as marine alkaloids, such as isoquinolinoid. The basic structural feature is the isoquinoline nucleus. These alkaloids have a wide range of medicinal properties, including antioxidants, antiviral, anticancer, antifungal, antispasmodic, and enzyme inhibitors. The most well-known isoquinoline alkaloids are morphine and codeine. The tyrosine and phenylalanine are the main source of their derivation. They are principally made from a precursor of dopamine (3,4-dihydroxytryptamine) linked with aldehyde or ketone. They are further divided into following catagories: 1- simple benzylisoquinoline alkaloids, for example reticuline and imbricatine; 2- isoquinoline alkaloids, for example salsoline and mimosamycin; 3- bisbenzylisoquinoline alkaloids, for example fumaricine; 4- pseudobenzylisoquinoline alkaloids, for example polycarpine and ledecorine; 5- secobisbenzylisoquinoline alkaloids, for example baluchistanamine; 6- isbenzylisoquinoline alkaloids with one ether link, for example dauricine; 7- manzamine alkaloids, for example manzamine a; 8- isbenzylisoquinoline alkaloids with two ether links, for example berbamine; 9- bisbenzylisoquinoline alkaloids with only aryl links, for example pisopowetine; 10- benzylisoquinoline alkaloids with one aromatic link and one or two ether links, for example rodiasine.

Indole Alkaloids

Indole alkaloids are the tryptophan's derived alkaloids and is the largest group of this type. These alkaloids include; simple tryptamine derivative, typtamine with one or more than one prenyl residues, carbazoles (ethanamine chain is losted), and some others in which incorporation of regular mono and diterpenoids occurs. When the source of these alkaloids is changed from terrestrial to marine, the structures also change. The common characterstics of these alkaloids is polyhalogenation. They are further divided into the following categories: 1- simple indole alkaloids, for example, aplysinopsin, gramine; 2- simple tryptamine alkaloids, for example, tryptamine; 3- bisindoles, for example, indirubin, 6,60 dibromoindigotin; 4-

quinazolinocarbazole alkaloids, for example, rutaecarpine; 5- cyclotryptamine alkaloids, for example, physostigmine; 6- carbazole alkaloids, for example ekeberginine; 7- β-carboline alkaloids, for example, harman; 8- ndolonapHthyridine alkaloids, for example, canthin-6-one; 9- ergot alkaloids, for example, ergotamine.

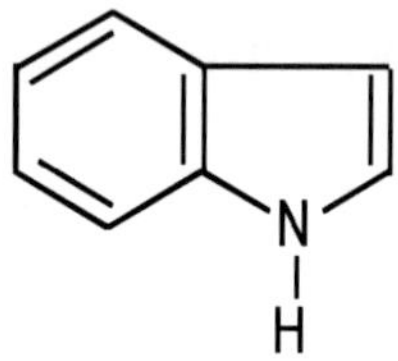

Steroidal Alkaloids

This type of alkaloid is distinguished by the ring system of 1,2-cyclopentane phenanthren. Typically they are present in higher plants, which belong to solanaceae, buxaceae, liliaceae, and apocynaceae families, some are also present in amphibians. There are many subtypes of these alkaloids, which are given as: 1- aminopregnanes, simplest type among all other types; 2- jerveratrum, for example, jervine); 3- salamandra, for example, cycloneosamandione; 4- solanidine, for example, rubijervine; 5- spirosolane, for example, soladulcidine; 6- conanine, for example, didymeline; 7- cerveratrum, for example, 3,6-cevanediol; 8- miscellaneous steroidal alkaloids, for example, bufotoxin; 9- buxus, for example cyclobuxine; 10- cephalostatins or ritterazines, for example, ritterazines a; 11- pregnane, for example, 20α-dimethylamino- 3β-senecioylamino-16β hydroxy-pregn-5-ene.

Imidazole Alkaloid

This type of alkaloid is distinguished by its imidazole ring structure. Since, during the precursor stage the imidazole ring is already formed, they are excluded from the structure transformation process. In this class, there are many examples from marine and microbial alkaloids which are structurally different. These alkaloids exhibit a diverse range of pharmacological activities as well as important pharmaceutical potential. The most important imidazole alkaloid is pilocarpine.

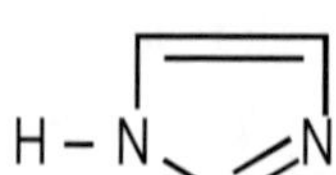

Purine Alkaloids

This alkaloid is the N-base (nitrogenous base) of nucleotide (DNA and RNA) or can say that it is the main constituent of DNA and RNA, as these nucleotides consist of purine ring, pentose sugar with pyrimidine base. The most typical examples of these alkaloids are theophylline, theobromine and caffeine. They originated from plants, and marine organisms as in the form of many of terpenoid-purine alkaloids, e.g., age lines and in the form of purines (substituted), e.g., phidolopin.

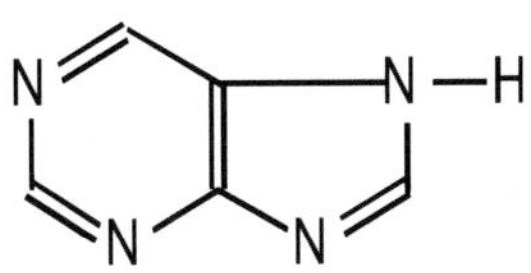

Pyrrolidine Alkaloids

These alkaloids consist of the basic pyrrolidine (C4N skeleton) nucleus. Many of these alkaloids are originated from plants. For example: ficine; the flavon nucleus and pyrolidine ring are involved, hygrine (biosynthesized from ornithine), and brevicolline; this is attached to the unit of β-carboline.

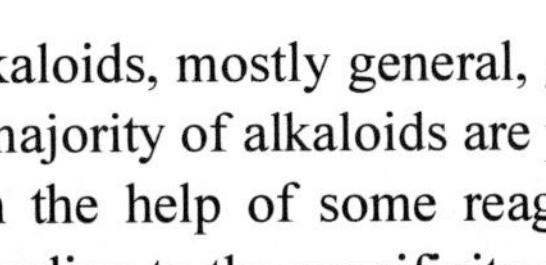

Tests of Alkaloids

In the identification of alkaloids, mostly general, group and specific reaction tests are performed. The majority of alkaloids are precipitated from neutral or mild acidic solution with the help of some reagents, which are named as precipitative reagents according to the specificity of reaction.

General Precipitative Reagents

1. Dragendorff reagent: is the solution of potassium bismuth iodide.
2. Wagner reagent: is the solution of iodine in potassium iodide.
3. Marme reagent: is the solution of cadmium iodide in potassium iodide.
4. Mayer reagent: is the solution of potassium mercury-iodide.

5. Sonnenschein reagent $H_3PO_4 \cdot 12MoO_3 \cdot 2H_2O$: Phospho-molybdic acid, a very sensitive reagent for alkaloid, because it gives amorphous precipitates a yellowish colour, which changes to blue or green with time (this is due to the reduction of molybdic acid).
6. Bertran reagent $SiO_2 \cdot 12WoO_3 \cdot 4H_2O$: contain silico-tungstic acid.
7. Schebler reagent $H_3\ PO_4 \cdot 12WoO_3 \cdot 2H_2O$: contains phospho-tungstic acid.
8. Hager reagent: is the saturated solution of picric acid.
9. 5% tannin solution.

At least 4-5 reagents are used for the identification of alkaloids because all these reagents do not necessarily give precipitate with all alkaloids.

In analytical way only negative results have some importance, because the reagents can also precipitated with protein and other N-containing compounds or substances.

Colour Tests

1. Concentrated nitric acid (HNO_3).
2. Concentrated sulphuric acid (H_2SO_4).
3. Wazicki reagent: it is the solution of *p*-dimethylaminobenzaldehyde in concentrated H_2SO_4.
4. Erdmann reagent: it is the mixture of concentrated HNO_3 and concentrated H_2SO_4.
5. Marquis reagent: it is the solution of formaldehyde in concentrated H_2SO_4.
6. Frede reagent: it is the solution of ammonia molybdate in concentrated H_2SO_4.

The "essential oil" term originated from a previously invented name, which is, "quinta essential." This name was first proposed in 16th century by Paracelsus von Hohenheim. They are secondary metabolites of aromatic plants which are made up of a variety of constituents. They are hydrophobic compounds with some aroma and have densities less than one. Various studies shows that thousands of aromatic plants, their leaves, flowers and roots synthesize the essential oils.

When the source of essential oils changes, their physicochemical properties are also changed, due to the changes in chemical composition. The Gas Chromatography-Mass Spectrometry (GC-MS) is the best analytical technique for analysis of chemical composition of essential oils because it provides quick results and precise data.

Analysis of Volatile Oil

In the analysis of volatile oil:

1. Addition of Sudan III solution in a small quantity of drug production of red colour in globules, confirms the presence of volatile oils.
2. Addition of a few drops of alkane tincture in a little quantity of drug, when red colour appears, the presence of volatile oil is confirmed.
3. Chemical test for mentha oil: Few drops of peppermint oil and 5 mL of HNO_3 solution (prepared by adding 1mL of HNO_3 to 300 mL of glacial acetic acid) is added and mix it well then heat it on water bath for 5 minutes until the liquid turns blue. Further heating deepens the colour, showing a copper colour fluorescence, and after further heating, a golden yellow colour is obtained.

Conclusion

Classical methods are widely employed for quantitative analysis to find the amount of each component present in a sample mixture. One or more than one chemical substance is analyzed in a classical analysis. One can deduce the identity of the analyte by observing the chemical reactions and their products. The chemical substances which are added are selected on the basis of their reaction with one chemical compound to form a distinctive chemical product. The first part of the chapter gave the basic concept about the acid-base, pH, buffer solution, pH measurement methods, oxidation-reduction reaction. The second part contains the classical methods. Classical methods can be categorized into volumetric analysis and gravimetric analysis. In these techniques, the complete chemical reaction between analyte and added substance is explained.

References

https://www.horiba.com/us/en/application/material-propertycharacterization/water-analysis/water-quality-electrochemistry-instrumentation/the-story-of-pH-and-water quality/the-story-of-pH/ways-of-measuring

pH/#:~:text=The%20methods%20for%20measuring%20pH,Glass%2Delectrode%20methods
https://onlinelibrary.wiley.com/doi/10.1002/9781119953647.ch5
https://byjus.com/chemistry/complexometric-titration/
https://acikders.ankara.edu.tr/pluginfile.php/134969/mod_resource/content/0/8th%20Week% 20Argentometric%20Titrations.pdf
https://facultystaff.richmond.edu/~rdominey/301/local/iodine_titrations.pdf
https://www.h2o2.com/technical-library/analytical-methods/default.aspx?pid=70&name=Iodometric-Titration
http://www.ecs.umass.edu/cee/reckhow/courses/572/572bk15/572BK15.html
https://www.lpdlabservices.co.uk/analytical_techniques/chemical_analysis/thermal_gravimetric_analysis.pH
https://www.sciencedirect.com/topics/chemistry/electrogravimetry
https://www.sciencedirect.com/topics/chemistry/saponification

Chapter 5

Electrochemical Methods of Analysis

Muhammd Asif[1], Malik Saadullah[2,*],
Bader Huwaimel[3], Jahan Ara Batool[2], Azhar Rasul[4],
Muhammad Uzair[5] and Bashir Ahmad Chaudhary[5]

[1]Faculty of Pharmacy, Islamia University of Bahawalpur, Pakistan
[2]Department of Pharmaceutical Chemistry, Government College University, Faisalabad, Pakistan
[3]Department of Pharmaceutical Chemistry, College of Pharmacy, University of Hail, Saudi Arabia
[4]Department of Zoology Government College University Faisalabad, Pakistan
[5]Faculty of Pharmacy, Bahauddin Zakariya University, Multan, Pakistan

Abstract

Electrochemical methods are used for the determination of analyte concentration or they measure the potential to characterize the analyte's chemical reactivity. Due to their high sensitivity, low analyte concentration requirement, high sensitivity, cost-effectiveness, portability, and ease of use, electrochemical techniques have acquired greater attention.

Keywords: electrochemical analysis, electrolysis, electrolytic, electrogravimetry, conductometry, coulometry, potentiometry, voltammetry, polarography, amperometric methods, oxidation, reduction

---

*Corresponding Author's Email: Maliksaadullah@gcuf.edu.pk.

In: Functional Analytical Techniques in Pharmaceutical Chemistry
Editor: Malik Saadullah
ISBN: 979-8-88697-059-3

Introduction

The chemical composition of a solution that forms part of an electrochemical cell determines the cell's potential, the current flowing through it, and its resistance. By measuring one or more of these electrical characteristics under controlled conditions, quantitative and qualitative data can be acquired. Direct measurements, in which sample solutions are compared to standards, can be done; alternatively, the electrical property changes during titration can be monitored to discover the equivalence point. Electrochemical methods are electrogravimetry, conductometry, coulometry, potentiometry and voltammetry. The solution's electrical properties are used in electrochemical methods to test analytes.

The method in which the electric potential of a particular electrode is measured in relation to analyte concentration is known as the potentiometric method. The methods in which electrolysis occurs and its product is measured are known as coulometric method or electrogravimetric method. The methods based on the dependency of current voltage are known as voltammetry and polarography. The method based on the measurement of current intensity during the reaction is known as the amperometric method. All electrochemical methods rely on the electric conductivity measurement of a solution.

Electrochemical Cell

A pair of electrodes in contact with the solution of electrolyte and faradaic current flowing through its electrode is referred to as an electrochemical cell. Galvanic cells and electrolytic cells are types of electrochemical cells.

Notation to Describe the Electrochemical Cell

The following universally accepted conventions are used to describe the structure of electrochemical cells.

1. On the right-hand side, the positive electrode (cathode), and on the left-hand side, the negative electrode (anode) is written.

2. The cell's anode is generally expressed by writing the metal first, followed by the metal ion present in the electrolytic solution, both are separated by semicolon symbol or vertical line.
3. The cell's cathode is generally expressed by writing the cation of the electrolyte first, followed by the metal present in the electrolytic solution, both are separated by semicolon symbol or vertical line.
4. Two parallel vertical lines are presenting the salt bridge that divides the two half cells.
5. The vertical line or single slash (/) denotes the boundary of phase.
6. The double vertical line or double slash (//) denotes the boundary of phase, whose potential is considered to be a negligible element of the overall potential of the cell. Boundary whose potential is considered to be an insignificant element of the total cell potential.
7. In the same phase, two components are separated by a comma (,).

Galvanic or Voltaic Cell

It is an electrochemical cell that generates current spontaneously when the electrodes are externally connected to a conductor (Figure 5.1).

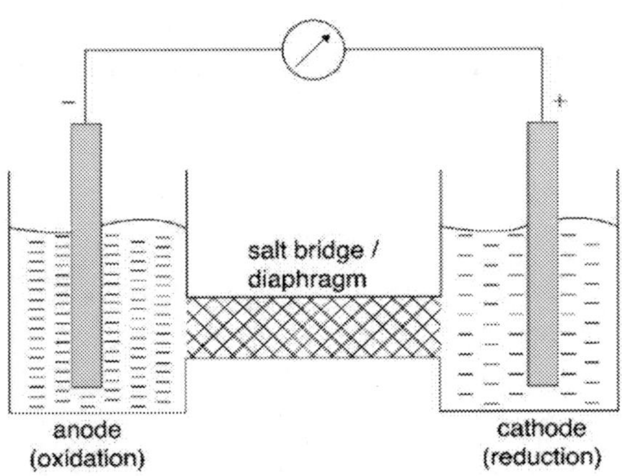

Figure 5.1. Schematic diagram of a galvanic cell.

Galvanic Cell Characterized by Following Features

1. In galvanic cells, spontaneous electrochemical reaction takes place when electrodes are externally connected by a conductor.
2. As a result, chemical energy is converted into electrical energy in a galvanic cell.
3. In this type of cell, the anode is a negative electrode and the cathode is a positive electrode.
4. In a galvanic cell, reduction takes place at positive electrode (cathode) and oxidation takes place at negative electrode (anode), after this, negative charges moved from cathode to anode within the cell, whereas negative charges moved from anode to cathode in the external circuit.
5. The primary cells (non-rechargeable cells), secondary cells (rechargeable cells), and fuel cells are the three types of galvanic cells.
6. A charged Pb-PbO_2 storage battery is an example of a galvanic cell.

Electrolytic Cell or Electrolysis

An electrochemical cell is one in which current is forced by a battery. It is commonly used to breakdown chemical compounds in a process known as electrolysis (Figure **5.2**).

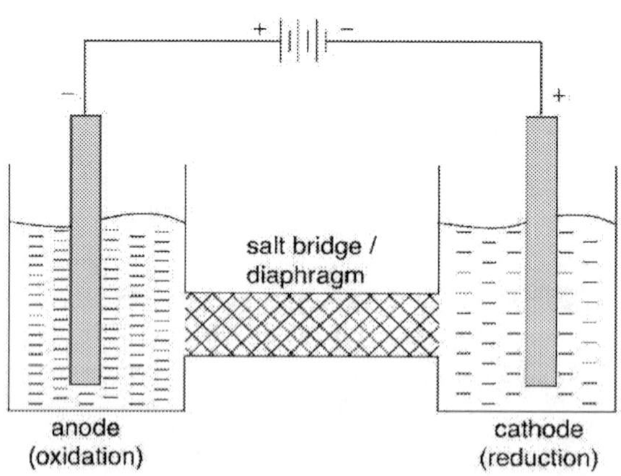

Figure 5.2. Schematic diagram of an electrolytic cell.

Electrolytic Cells Are Characterized by the Following Features

1. Electrochemical reaction in electrolytic cells is triggered by an externally provided voltage larger than the cell's open-circuit voltage.
2. This results in the conversion of electrical energy into chemical energy.
3. In these type of cells, the anode is a positive electrode and cathode is a negative electrode.
4. In an electrolytic cell, reduction takes place at the positive electrode (cathode) and oxidation at the negative electrode (anode), after this, negative charges moved from cathode to anode within the cell, whereas negative charges moved from anode to cathode in the external circuit.
5. These types of cells are employed in a variety of technical procedures, including electrorefining of metals, electroplating and electrolytic synthesis.
6. A charged Pb-PbO_2 storage battery is an example of electrolytic cell.

Half-cell reactions: Oxidation or reduction reaction takes place at an electrode.

Cathode: The electrode where oxidation takes place.

Anode: The electrode where reduction takes place.

Reversible Cell

If half-cell, the reactions are reversed by the reversal of the current flow, that cell is considered to be thermodynamically in equilibrium.

Standard Hydrogen Electrode (SHE)

It is used as a primary standard electrode, to which the potential of all electrodes is referred and used for high precision measurements in cells without liquid junction. The hydrogen electrode potential is adjusted to the standard atmospheric pressure of 1. The SHE comprises a platinum wire, which is electroplated with platinum salt, on which H_2 is bubbled.

Standard Electrode Potential

In a solution, the electrode potential is monitored. Current and potential is measured electroanalyticaly, which is not dependent on the composition of the solution, while it totally relies on the response of one electrode. SilverÀWA chloride and calomel used as reference electrode.

Ohmic Drop (IR)

A potential develops in an electrochemical cell when current (I) flows, it is due to the cell resistance (R) and it is determined by the product (IR). It is always subtracted from the theoretical cell potential, lowering the potential of a galvanic cell and increasing the potential needed to operate.

Electrogravimetry

Although the chemical effect of electric current was noticed in the early 1800s immediately after inventing the Volta column battery, analytical applications of the phenomena required years to develop.

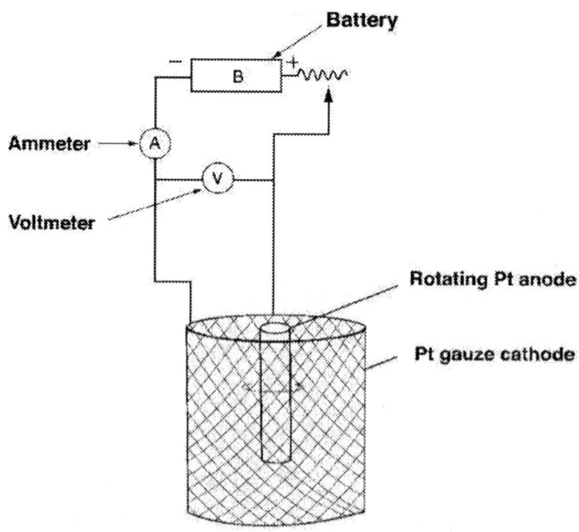

Figure 5.3. Schematic presentation of electrogravimetry.

Wolcott Gibbs weighed the amounts of metal deposited on the mercury cathode and, in 1864, he was the first founder to use copper via electrogravimetric technique (Figure 5.3). In 1882, Classen published the first book on electrogravimetry. The electrogravimetry method is moderately sensitive, fairly fast and has a precision of several parts per thousand for a broad variety of metals, present individually or in mixture. Electrogravimetry is a technique in which a metal is basically quantitatively electroplated onto an electrode, usually platinum. The amount of metal plating is determined by the difference in the electrode weight before and after electroplating.

Principle

The basic principle of the electrogravimetry method is that in the electroplating procedure, a current is forced to flow between a platinum gauze cathode and a platinum wire anode using a circuit. The cathode is removed, dried, and weighed after complete metal deposition on it. The other two objectives of electrolysis are the following:

(1) Electrolytic deposition to achieve a separation of the deposited species from the rest of the solution (electro-separation).
(2) Electrolytic deposition for preparative purposes (electroplating).

On the electrode, a metal is electrolytically deposited by enhancing the electrode mass.

$$E_{electrolysis} = E_{cathod} - E_{anode}$$

The quantity of electrons deposited on the electrode is directly proportional to the amount of current passed through the solution, and the quantity of different substances deposited is directly proportional to the molar mass divided by the number of electrons involved in the process of electrolysis, according to Faraday and Ohm's laws of electrolysis.

Electrochemical Cell Construction and Working

The electrochemical cell consists of a glass vessel. The glass vessel is filled with the solution of electrolyte, and the electrodes are immersed in it. The cell is covered externally with an isolated glass wall jacket linked to a thermostat, to maintain a constant electrolyte temperature (Figure 5.4). Generally, three different electrodes are utilized in electrochemical experiments: working, reference, and auxiliary electrodes. The WE (working electrode) is the substrate conductor on which the chemical reaction of interest occurs, and the potential of a working electrode is monitored with respect to the reference electrode. An auxiliary electrode, also known as a counter electrode, is required for the flow of ionic current; it closes the system's electrical circuit and permits current to flow from the external system. The transport process of species at the electrode interface may also be maintained with the magnetic stirrer. A saturator is used to remove the dissolved O_2 within electrolytes, and maintain the inert gaseous phase within the cell. These cells are usually employed in the electrodeposition processes, in which the electrode ions are deposited across the surface of the substrates by a reduction process.

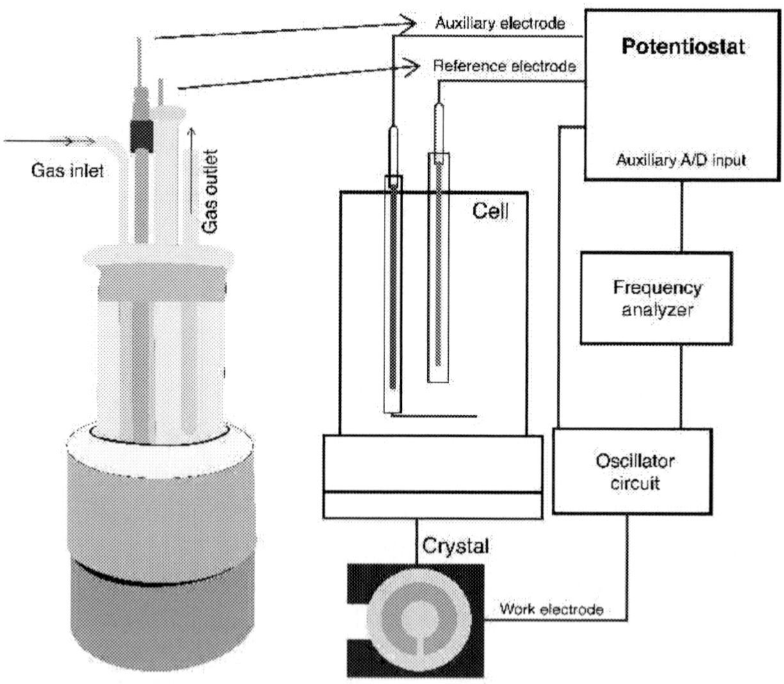

Figure 5.4. Experimental configuration of a conventional three-electrode electrochemical cell used in measurements coupled to the electrochemical quartz–crystal microbalance (EQCM).

Different Cell Design

There are three key distinctions between a cell design on the basis of electrode surface area, stirring, and electrode separation in bulk electrolysis and voltammetric cells.

Surface Area

In a voltammetric experiment, a small fraction of redox active species needs to be electrolyzed, so the diameter of the working electrode ranges from micrometers to a few millimeters. In bulk electrolysis, on the other hand, fast electrolysis occurs, which uses a large-surface-area working electrode.

Stirring

In voltammetry experiments, no stirring is required. But, in bulk electrolysis experiments, the solution needs to be stirred.

Electrode

The working electrode may be a mercury pool, glassy carbon, platinum gauze, or other form of carbon. An auxiliary electrode with larger surface area is required, since large currents are generated. Electrolysis takes place at the auxiliary electrode during electrochemical analysis. But, the duration of voltammetric experiment is generally short enough that the products of electrolysis at the auxiliary electrode do not have time to be dispersed to working electrode.

Types of Electrochemical Methods

Bulk electrolysis (BE) is a method also known as constant current (galvanostatic) or constant potential (potentiostatic) electrolysis.

Constant Current Electrolysis

In galvanostatic or constant current electrolysis, as the name indicates, the current (I) is kept constant, and potential is increased. A fixed current amount can be passed between the cathode and anode. It cannot be used for the separation of ions in a solution containing single species, which is the limitation of constant current electrolysis (Figure 5.5).

Factors Affecting the Deposition

1. Current density
2. Temperature
3. Chemical nature of ion

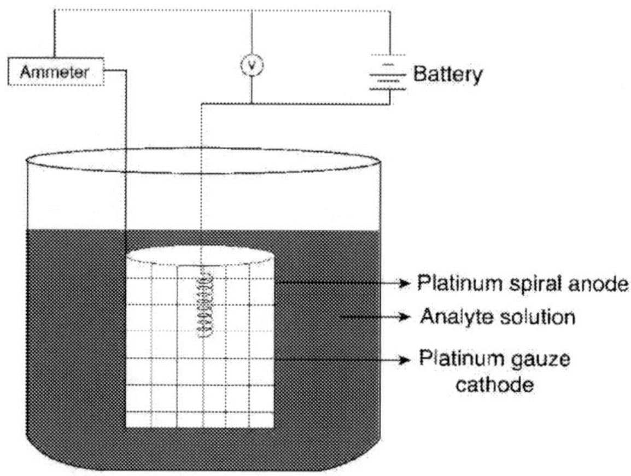

Figure 5.5. Schematic diagram of constant current electrogravimeter.

Constant Potential Electrolysis

In potentiostatic or constant potential electrolysis, as the name indicates, the potential (P) is kept constant, and current is increased. Until the solution is entirely electrolyzed, the WE is kept at a constant P. For instance, the electrolysis potential is generally chosen to be slightly more positive than the E1/2 value of a species being oxidized or slightly more negative than the E1/2 value of a species being reduced, and therefore just into the limiting current region of a steady state voltammogram (Figure 5.6). It's more commonly used to separate the components from a mixture in which the decomposition potentials are not widely separated.

Application of Electrogravimetry

1. Electrogravimetric technique is used to examine ion and solvent transport across the conducting polymer such as polyaniline.
2. Using ac quartz electrogravimetry to investigate the formation and dissolution of electrochemical phase.

3. This method is used to remove the trace metals from the sample.
4. Electro-synthesis by using the electrogravimetry technique.
5. Spontaneous metals deposition is done by the electrogravimetry technique.
6. Electrogravimetry technique is especially used in the successive deposition of the metals.

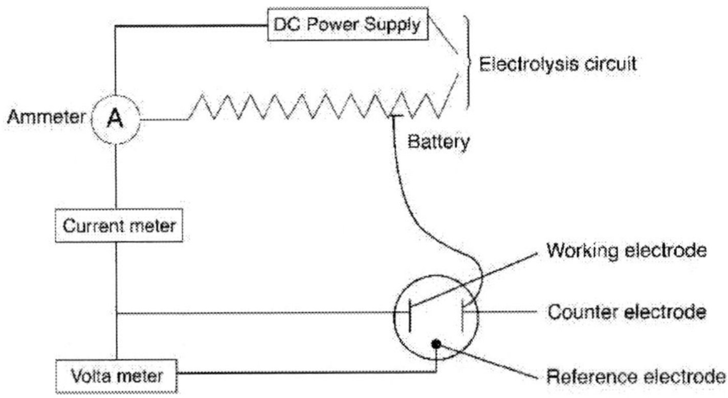

Figure 5.6. Schematic diagram of constant potential electrogravimeter.

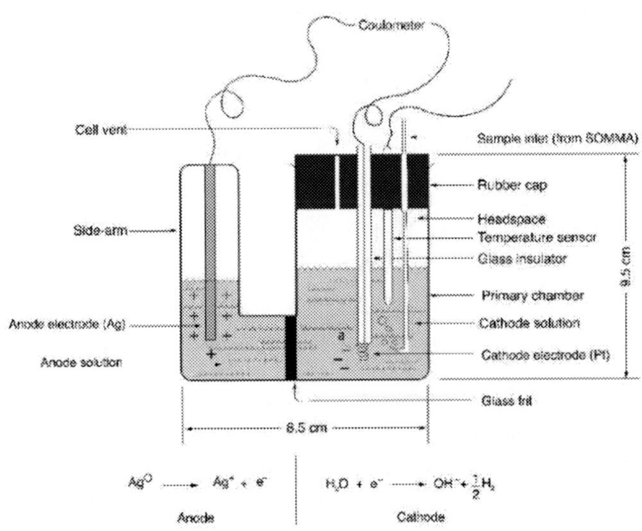

Figure 5.7. Schematic presentation which shows typical arrangement for a coulometric sensor made of a cathodic and anodic half reactions for the coulometric titration of the H^+ from the acid formed by the reaction of CO_2 and ethanolamine.

Coulometry

This is an electrochemical technique for estimating the total charge consumed during the redox conversion of an analyte at an electrode. Coulometry is usually an absolute method, which means it does not require calibration because the electrical charge can be determined with high precision (Figure 5.7). Coulometry also has the benefit of being inexpensive due to its inherent simplicity. This method was popular in the mid-twentieth century, although voltametric alternatives such as differential pulse polarography have since supplanted it in many applications. But several applications are still in use today, most notable the Karl Fischer technique used for determining the water content. In 1834, Michael Faraday described two basic rules of electrolysis. In fact, Faraday's Laws are also based on two fundamental laws, 1) conservation of matter, and 2) conservation of charge. According to the first law of Faraday, the amount of deposited material (m) during electrolysis is directly proportional to the current (I) and time (T). According to the second law, the equivalent mass of material to be electrolyzed determines the amount of product that produced during electrochemical reaction.

Principle and Theory

Coulometry measures the total current (I), either directly or indirectly, to estimate the number of electrons passed by entirely converting an analyte from one oxidation state to another at the WE. We can measure the concentration of analyte based on the number of electrons passed, by using Faraday's equation:

$$m = \frac{M}{nF}Q = \frac{MIt}{nF}$$

In the above equation, **Q** is the consumed amount of charges, **n** is the number of charges, **I** is the amount of current, and **t** is the time duration of electrolysis process. The measurement of charges is a good tool for both quantitative and qualitative analysis, if the current efficiency is 100 percent, it means the total current is consumed by electrode reaction. The amount of the substance in the solution can be determined by knowing M and n. The estimation of electron number (n) transported during the process of electrode is critical in many situations, notably in organic electrochemistry, for elucidating the reaction mechanism. The entire amount of charge required for complete electrolysis of a known amount of substrate must be calculated for this purpose.

Types of Coulometry

There are two types of coulometry, which are, controlled-potential and controlled-current coulometries.

Coulometry at Constant Potential

The total amount of charge (Q) can be measured directly using a coulometer or by integrating the current (I) or time (t) curve. The principle of coulometry at constant potential, the end point I=0, i.e, when the concentration of species under examination reaches zero, can only be achieved over an indefinite period of time; however, in reality, the electrolysis process typically stops after the current has fallen to a few percent of its original value. For a stirred solution and simple electrolysis, the change in current (I) and charge (Q) as a function of time at a constant potential E>Ee is as follow:

$$I(t) = I(t = 0)exp\left(-\frac{DA}{\delta V}t\right)$$

And

$$Q(t) = Q(t \rightarrow \infty)\left[1 - exp\left(-\frac{DA}{\delta V}t\right)\right]$$

In the above equation, **D** is the diffusion coefficient of reactive species, **A** is the area of electrode, **d** is the diffusion layer thickness, and **V** is the solution volume. The E (applied potential) is basically far from the respective Ee (equilibrium potential). It is also feasible to create a reactant in a predetermined amount by electrolysis, which subsequently enters a reaction with a solution component. In coulometric titration, it is used to detect the endpoint in an ordinary way, for example by use of an indicator. For both liquid and solid electrolytes, gas titration, a kind of coulometric titration, is utilized.

Controlled-Current Coulometry

It is a more complex, but faster technique. The charge used during the reaction is directly proportional to the time period of electrolysis process, which is one of its biggest advantages.

Three electrodes are used in controlled-potential coulometry to set the potential. The liquid mercury electrode or a platinum electrode is used as the working electrode. A salt bridge separates the auxiliary electrode from the analytical solution, which is typically composed of platinum. The reference electrode is usually a saturated calomel or silver chloride electrode. The potentiostat monitors the current as a time function for measuring total charge (Q) and utilizes electronic integration as time function for calculating the associated charge. The analyte concentration in the solution drops during the electrolysis process, resulting in a decrease in electrolysis rate. This leads to a further increase in the analysis time, which might be difficult for industrial applications. Constant current during electrolysis which shortens analysis time (e.g., <110 min) has the main benefit of the controlled current coulometry approach. Moreover, the estimation of analyte concentration is simple, as the charge (Q) is only a product between the constant current and time. Nevertheless, it is difficult to assure a 100% current efficiency and to know when the electrolysis process will be complete and if refining is necessary in the measuring process. Controlled-current coulometry is normally done by using a constant current during the electrolysis process, comprised of platinum-based counter electrode (CE) and working electrode (WE). This counter electrode can be separated from the analyte by using a salt bridge. In order to properly measure the electrolysis time, this technique also requires an accurate clock. Coulometry determined the concentration of the lowest analyte depending on the capability to correctly estimate the endpoint of electrolysis. 0.1% to 0.3% accuracy is achieved, when using controlled-current coulometry, whereas, 0.5% accuracy is achieved in controlled-potential coulometry.

Applications of CPC and CCC

1. The concentration of more than 50 chemical elements, including trace metals and halide ions, has been estimated by CPC analysis. CPC is commonly employed in oxygen measurement and in uranium and plutonium determination.

2. CCC is more fruitful and versatile, therefore more widely used than CPC. CCC can estimate the thickness of the metallic coating and determine the water concentration in the order of mg/L in commodities like sugar, cheese, butter, paper, and petroleum. For a variety of analytes that may be measured by redox titrimetric such as acid-base, complexation, or precipitation, CCC methods have been devised.

Coulometric Titration

There is a direct application of Faraday's laws of electrolysis in coulometric titration, which was first used in this form, in 1939, by Somogyi and Szebelledy.

Applications

1. The quantitative analysis of different organic and inorganic compounds have been done by using coulometric titration.
2. Coulometric titration is a more versatile method to determine the carbonate content of calcium phosphate samples.
3. The coulometry method is used to determine the metallic coatings thickness by measuring the amount of current required to dissolve the coating.
4. The coulometry titration is used to estimate the water concentration in various samples on the order of mg/L.
5. Coulometry technique is used in the determination of the picric acid.
6. The separation of nickel and cobalt is achieved by coulometry.
7. Used in the determination of n-values of the organic compounds.
8. Used in the determination of the environmental pollutant.

Conductometry

Conductivity or conductance measurements in conductometry are related to a solution's capacity to transport current when a steady alternating current potential is maintained in a conductivity cell (Figure 5.8).

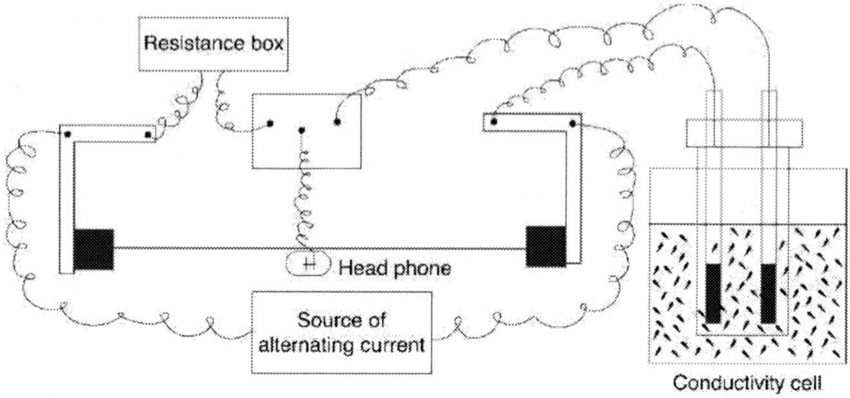

Figure 5.8. Schematic presentation of conductometry.

Principle

The electrolyte concentration in aqueous solutions is measured by using the conductometry technique. The measured electrical resistance of solution can be used to estimate the molarity of analyte that generates solution conductivity. The formula for determining the dependency of resistance concentration is:

$$C = \mathrm{Const}_{\mathrm{cell}} \frac{1}{\Lambda_m} \frac{1}{\mathrm{Res}}$$

where $\mathbf{Const_{cell}}$ is the constant that basically dependent on the measuring cell, **Res** is the electrical resistance measured with the instrument (from Ohm's law **Res** = **I/V** and at constant voltage **V** the intensity **I** measurement allows the calculation of *Res*), and $\mathbf{\Lambda_m}$ is the equivalent conductivity for the ionic species. Although $\mathbf{\Lambda_m}$ can be taken for practical purposes as constant, but it varies with the concentration following Kohlrausch's law:

$$\Lambda_m = \Lambda_m^0 - \Theta\sqrt{c}$$

Where $\mathbf{\Theta}$ is a constant and $\mathbf{\Lambda_m^0}$ is the limiting molar conductivity which is specific for each ion. Molar conductivity is temperature-dependent.

Method

At a steady temperature, the solution of sample is put in the conductivity cell. With the help of the thermostat a constant temperature is maintained. The cell is then linked to the resistor box (R), and alternating current is routed through the cell with the assistance of the induction coil. The conductivity of solution is then measured by using the equation:

Conductivity of the solution = 1/resistance of the solution

Applications

1. Serjeant and Albert presented a method for determining pKa by conductivity measurements. Conductometry was employed to determine the pKa of lidocaine in recent years. The result obtained was 0.7 unit lower than the accepted pH-metric value.
2. During enzyme processes, continuous recording was carefully examined and estimated using the conductometric technique.
3. Ammonia (NH3) is a basic gas which can also be determined by conductometric technique.
4. Conductivity is a universal method for the ion's detection.
5. Overall ionic content of the aqueous solution can be determined by conductometry technique.
6. In order to study the electrical characteristics of ions like migration, transport, and diffusion, conductometric techniques are used.
7. Conductometric titration can enhance the accuracy of end point detection and can provide analysis of weak electrolytes, e.g., boric acid.
8. Used in the quantitative analysis of compound.
9. Used in the determination of water purity.
10. Used to determine sea water salinity.

Voltammetry

Jaroslav Heyrovsky, in 1922, invented this technique for the first time, for which he won the Noble Prize in chemistry in 1959. Various materials for working electrodes are available, such as carbon, mercury, platinum, silver, and gold. The auxiliary electrode is usually constructed of platinum, hydrogen, calomel, silver, and silver chloride electrodes that can be used as reference electrodes. Voltammetry is an electrochemical technique in which the electrical current is measured over a period of time as a function of changing potential. Voltammetry is a technique that involves applying a time-dependent potential to working electrode and measuring the current that flows between auxiliary electrode and working electrode as a function of that potential. The voltammogram is a graph of current vs. applied potential. There are several potential excitations in controlled potential techniques, such as sine wave, potential step, ramp, and pulse trains; as a result, voltametric techniques can be applied for both qualitative and quantitative analysis (Figure 5.9).

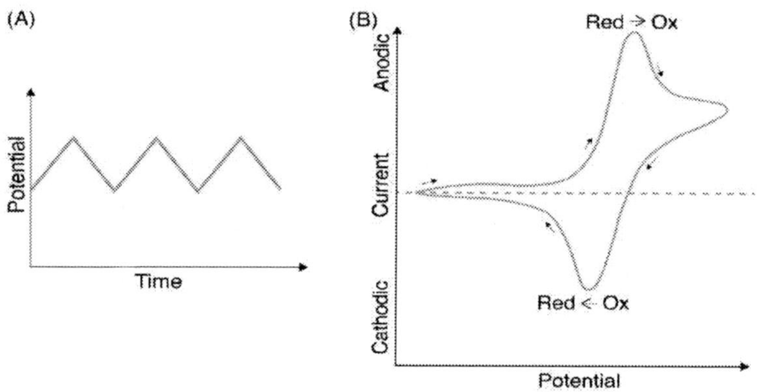

Figure 5.9. Schematic presentation of Potential-time excitation signal in a CV experiment in (A), and a cyclic voltammogram for a reversible $\mathbf{Ox} + \boldsymbol{ne^-} \rightleftharpoons \mathrm{Red}$ reaction in (B).

Types of Voltammetry Methods

Hydrodynamic Voltammetry

Inducing convection with the aid of stirring is known as hydrodynamic voltammetry. In this technique, current is calculated as a function of the voltage supplied to a working electrode. In hydrodynamic voltammetry, the applied potential profile is the same as in polarography, either a differential pulse or a linear scan and the resultant voltammogram is also identical to those in polarography, with the exception of the lack of current oscillation caused by mercury drop growth. Moreover, this technique is not limited to Hg electrodes.

Pulse Voltammetry

Pulse voltametric technique was first developed by Jenkins and Barker in 1952. The majority of pulse voltametric methods rely on current or potential measurement of the sample. The current sampling regime and excitation waveform are the key differences between these methods.

Like polarography, pulse voltammetry techniques can be classified in four categories:

1. Normal-pulse voltammetry (NPV)
2. Reverse-pulse voltammetry (RPV)
3. Differential pulse voltammetry (DPV)
4. Square-wave voltammetry (SWV)

The current is measured twice in DPV, first before the pulse application (t1) and again later in the pulse life (t2). The difference of current is represented as [$\Delta i = i(t2) - i(t1)$] and is instrumently plotted against the applied potential. The height of the current peak in the resultant differential pulse voltammogram is directly proportional to the corresponding analyte concentration. The DPV method is a highly useful approach for the measurement of trace amounts of inorganic and organic compounds. It also gives information about the chemical form of the analyte such as complexation and oxidation states, analyze the sample of concentration as low as 1028 M and also provide information about the chemical form of analyte (oxidation states, complexation, etc.). SWV is a large-amplitude differential voltammetry. The major advantage of SWV is its extremely fast speed, moreover, in SWV, both forward and reverse currents are measured, therefore the current increases four times higher than that of DPV leading to an increase in sensitivity. The most important benefit of SWV is its fast speed.

Stripping Voltammetry

In this technique, the current is primarily measured as a function of applied potential, whereas the analyte is first deposited at the electrode surface and then electrochemically removed or stripped. Generally, in stripping method, two steps are involved, one is deposition and the other is stripping. In the first step, known as deposition, the solution contains metal ions, their small amount is deposited on the electrode. In the second step, known as stripping, the electrochemically dissolution of deposited metal ions takes place from the surface of the electrode.

Cyclic Voltammetry

It is the most commonly used method for the acquisition of quantitative information concerning electrochemical processes. In cyclic voltammetry, a triangular potential waveform is applied to linearly scan the potential of WE. The resultant current was monitored throughout the potential sweep and the resultant cyclic voltammogram was plotted between current and potential. In electroanalytical studies, CV is usually the first experiment. Depending on the information needed, a single or many cycles can be employed. Due to the current peaks, it is useful for quantitative analysis.

The choice of voltammetric method depends on the nature of the sample. The main advantage of voltammetry technique is that it quantitatively analyzed more than one analyte in a single specimen. Concurrent analysis may be performed when voltammogram for two components overlap.

Applications

1. The voltammetry method is most widely used generally for the analysis of chlorine, dissolved O_2 and metals in the environmental sample.
2. Voltammetry methods, such stripping voltammetry and anodic voltammetry, which are frequently combined with polarographic techniques, to measure the metals.
3. Trace metals can be analyzed in the environmental sample (e.g., sea water, river and ground water) by using voltammetry techniques.
4. Voltammetry methods used to measure the concentration of carbon dioxide, ammonia, and sulfuric acid in atmospheric air, and amount of dissolved oxygen and anionic surfactant in waters and waste water.
5. The amount of trace metals in the blood and urine sample can be measured by voltammetry techniques.
6. Other compounds are detected in clinical samples by using voltammetry methods, such as glucose, ethanol, formaldehyde, lactate, and choline.
7. It is used to figure out what is the kinetic rate of reactions.
8. Thermodynamic properties can be determined by using this method.

Principle Methods Involved in Voltammetry

In voltammetry, there are two main techniques.

1. Polarography: Polarography is the measuring of current at various potentials.
2. Amperometry: Amperometry is the measuring of current at a constant potential.

Polarography

It is a voltammetry technique, in which the deposited ions or molecules at dropping mercury electrode are oxidized or reduced at an applied potential (Figure 5.12). The first voltametric technique was polarography. In 1922, Heyrovsky developed polarography. He defines polarography as electrolysis via polarized DME. In 1925, Shikata and Heyrovsky constructed a polarograph, which automatically recorded current-potential curves. In 1959 both were awarded the Nobel Prize in the field of chemistry. In the field of analytical chemistry, it was the first automatic apparatus. Moreover, additional development in the method is associated with the names of Barker (square wave and pulse wave), Matheson and Nichols (oscillopolarography), and Heyrovsky (derivative polarography).

Principle

The basic principle of polarography is the analysis of solution or electrode processes by electrolysis with both polarized and unpolarized electrodes. It is a form of measurement that falls under the category of linear sweep voltammetry, in which electrode potential is altered in a linear fashion from start to end. The voltammograms of the normal polarography are generated by a linear scan of the potential. Although polarography is performed in an unstirred solution, falling mercury droplets produce current and current oscillation caused by mercury drop growth, which may induce a time-dependent change in the area of WE. The main advantages of an electrode is its wide cathodic potential range and renewable surface, because of unique characteristics of this electrode, polarography has been widely used to determine several important reducible species.

Types of Polarography

Normal polarography has been replaced by different types of pulse polarography, which are the following:

1. Differential pulse polarography (DPP)
2. Normal pulse polarography (NPP)
3. Staircase polarography (SCP) and
4. Square-wave polarography (SWP)

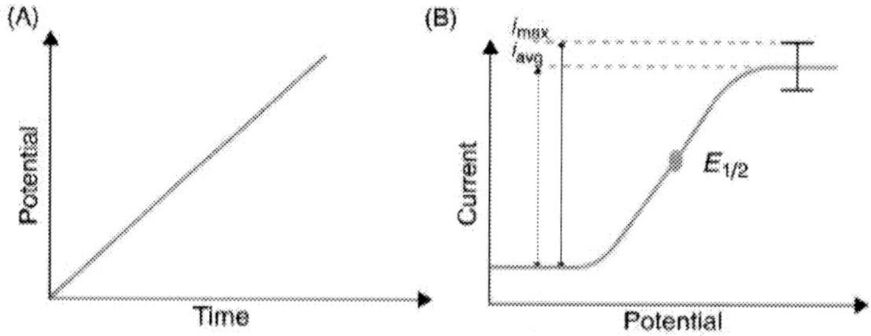

Figure 5.10. Schematic diagram of Potential-time profile in (A) and voltammogram of normal polarography in (B).

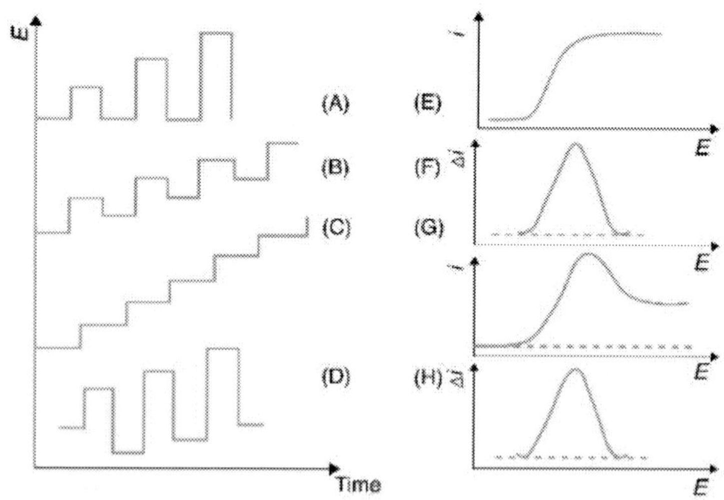

Figure 5.11. Schematic diagram of Potential-time profile and voltammograms for (A, E) normal pulse, (B, F) differential pulse, (C, G) staircase, and (D, H) square-wave polarography.

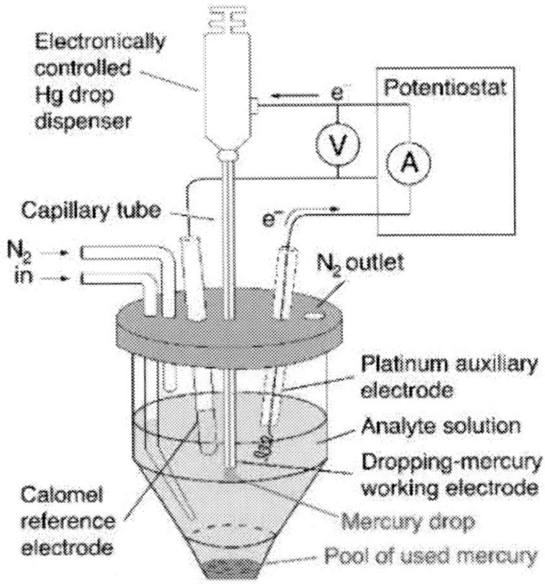

Figure 5.12. Schematic presentation of polarography.

In these cases, the limiting and peak currents are directly related to the concentration of analyte. Furthermore, quantitative analysis may be done with half-wave and peak potentials. Pulse polarography has better sensitivity and detection limits than normal polarography (Figure 5.10 & Figure 5.11).

Apparatus and the Methods

Their apparatus, comprised of a DME, serves as both cathode and WE. The anode is a pool of mercury at the bottom of a reservoir which serves as a reference electrode (RE), with a constant potential. Both electrodes, the cathode and anode are parallelly placed in the solution of sample. After this, both electrodes are connected with a battery, voltammeter, and galvanometer. Then constant voltage is applied with the aid of recorder record the current-voltage graph. The sample cell is constructed of glass with a curvature edge on which to put the mercury. The cathode capillary is immersed into the solution of sample by setting the drop duration of 2 to 7 seconds. The electrolytes that support them, like saturated KCl, are employed for controlling the ion movements at the electrode surface. The alkaline pyrogallol solution generally removes the O_2 in the solution of sample. The diffusion current obtained is directly proportional to the amount of sample solution.

Electrodes

In polarimetry, the electrochemical cell is mainly composed of following three electrodes.

1. Working electrodes: The WE are employed to access the response of analyte upon applying potential, e.g DME
2. Dropping mercury electrode: First, Barker introduced DME. The main concept of this electrode is to regulate the flow of mercury through capillary tube, closed by needle valve. It is applicable to +0.4 to −1.8 V, is the primary benefit of this electrode.
3. Auxiliary electrode: The circuit between potentiostat and WE is completed with this electrode.

Limitations of Polarography

To access the activity of analyte using polarography has many drawbacks

1. For direct ultra-trace elemental analysis this technique is not sensitive enough.
2. A long time for analysis (minutes compared to seconds for other analytical techniques such as AAS, ICP-AES) is required.
3. More critical operator experience (understanding of electrode processes is essential), in data interpretation.
4. Limited elemental range owing to the constraints of the voltage window of the dropping mercury electrode.
5. Toxic nature of mercury has recently limited its use.

Advantages

1. Comparatively required inexpensive equipment.
2. Ability of the polarography technique is to differentiate between elemental oxidation states (i.e., Cr, As).
3. Ability of the polarography technique is to establish the chemical form of elements (e.g., characterization and quantification of metal-organic complexation in natural waters).
4. Polarography provides simultaneous elemental analysis (e.g., Cu, Pb, Cd, Zn).
5. Generally, sensitivity down to 10-8 mol l^{-1}.

Applications

1. Used in the determination of the local anesthetics.
2. Used in the functional group analysis.
3. Used to determine the vitamin C in the food beverages.
4. Used in the determination of the composition of alloys.
5. Used in the estimation of the trace metals like Zn, Fe, Mn, Cu.
6. By using differential pulse voltammetry (polarography) (DPV, DPP), adenine and guanine can be analyzed at a carbon electrode and cytosine at a mercury electrode.
7. Metal ions, organic compounds and inorganic ions such as IO^{-3} and NO^{-2} can be easily analyzed by polarography.
8. Organic polarography has a wide range of analytical and quality control applications. In a broad range of matrices such as pharmaceutical formulations (tablets, creams, ointments, and injections), biofluids and extracts, over a hundred of pharmaceutical compounds have been determined by polarographic technique.
9. The polarographic results, with an analytical precision of 1 to 5%, are better suited for the assessment of minor constituents of pharmaceutical formulation (w/w content <5%).
10. In the bulk pharmaceutical formulations, impurities can be detected by using polarographic method

Amperometry

The most sensitive technique is amperometry, but it is only responsive to electroactive analytes. The amperometry term is derived from the unit of current "ampere" and the measurement device used in this technique "meter". The amount of current recorded usually depends on the oxidized or reduced concentration of the substance, so that's why this process may be used for many analytical purposes. A constant voltage is supplied to the WE in the amperometric technique, measured the response of current as a function of time, and potential is not scanned as a result there will be no voltammogram.

Principle

In simple amperometry technique or in constant potential amperometry technique, to oxidize or reduce the analyte, adequate potential is applied to the electrode. In this situation, current integration generally yields a fair estimate of analyte concentration, according to Faraday's laws of electrolysis

$$N = Q/nF$$

In the above equation, **N** presents the substance concentration in moles, **Q** presents total charge, **n** presents total electrons number transferred per ion, **F** presents the Faraday constant.

Electrodes

In polarimetry, an electrochemical cell is mainly composed of following three electrodes.

Working electrodes: The WE are employed to access the response of analyte upon applying potential, e.g DME

Dropping mercury electrode: First, Barker introduced DME. The main concept of this electrode is to regulate the flow of mercury through capillary tube, closed by needle valve. It is applicable to +0.4 to –1.8 V, is the primary benefit of this electrode.

Auxiliary electrode: The circuit between potentiostat and WE is completed with this electrode.

Applications

1. To examine the many other types of substances in samples ranging from water to radioactive materials, amperometric titration commonly and extensively employed.
2. Amperometric biosensors and amperometric sensors are extensively used in environmental, industrial and clinical applications. Moreover, since the 1970s, amperometric techniques have been widely used in HPLC to detect different substances.
3. Sensor manufacturing is the most significant use of amperometric technique. In 1956, Clark developed the first amperometric sensor, it is used to estimate the dissolved oxygen in the blood.
4. Quantify the metal ions in the mixture by amperometry technique.
5. Water analysis done with Karl Fischer reagent by using amperometry technique.
6. Amperometry method also used in the analysis of alloys.
7. Development of amperometric biosensors might be done by using different bioreceptors, these amperometric biosensors are utilized to detect toxins and contaminants in foodstuff (Figure 5.13).

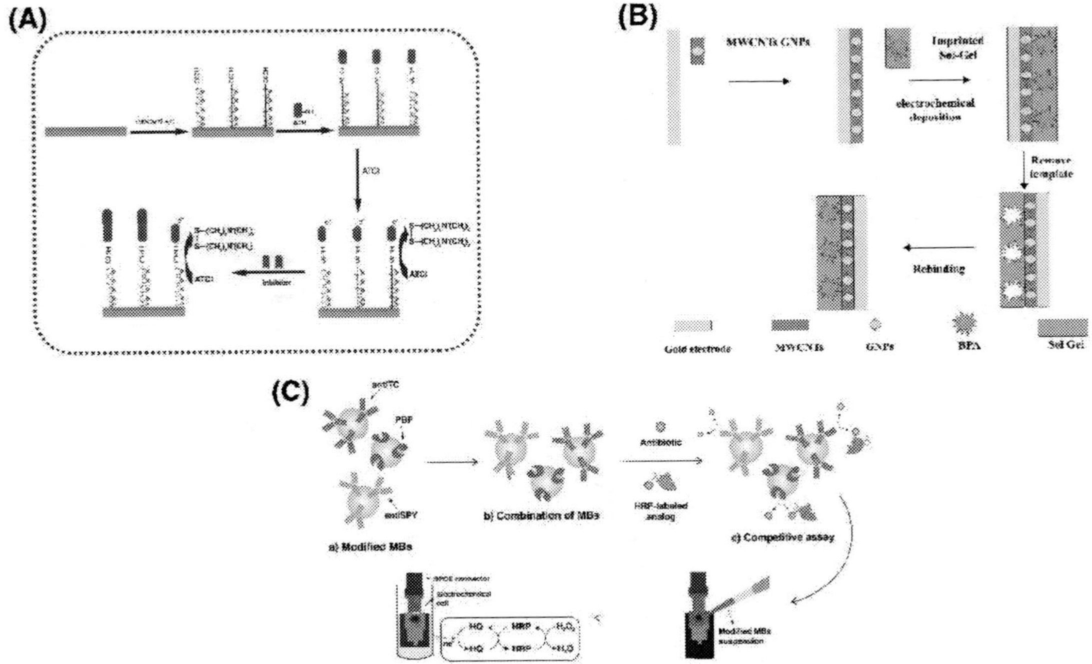

Figure 5.13. Schematic presentation in which three examples of amperometric biosensors developed for the food contaminants detection, based on (A) enzyme, (B) antibodies and binding protein, and (C) MIP. (A) Amperometric biosensor based on the inhibition of acetylcholinesterase (AchE) by OPPs. (B) Amperometric biosensor based on the specific recognition of bisphenol A (BPA) by a thin film of sol-gel molecular imprinted polymer. (C) Multiplex amperometric detection of antibiotics based on magnetic beads immobilized with antibodies and binding proteins.

Potentiometry

In 1983, the first potentiometric titration was carried out at the Leipzig Ostwald Institute by Behrend. He used a mercury electrode and mercury/mercury electrode to titrate mercury nitrate with standard solutions, either potassium bromide or potassium chloride. He used a capillary electrometer to measure the electromotive force. In 1893, Le Blanc discovered the hydrogen electrode, but he did not know its use for analytical purposes. In 1897, Bottger conducted acid-base titration by using hydrogen electrodes. In 1900, Crotogino was the first person to use potentiometry to redox titrations. He used a platinum indicator electrode to titrate the iodide with potassium permanganate. In the 1930s, after the introduction of electronic equipment commercially, it became a widely accepted analytical method. Nonetheless, some potentiometric techniques and endpoint detection approaches have been far older than usual and their principle developed around the beginning of the 20th century

Generally, in the potentiometry method, the indicator electrode potential is measured against the reference electrode. The indicator electrode is directly dipped into the analyte solution, whereas, a salt bridge is used to separate the reference electrode from the analyte solution. The indicator electrode potential is directly proportional to the logarithm of the analyte's activity in solution (Figure 5.14). To identify the end point of titration, potentiometric techniques have been and are still widely used. Ion- selective electrodes were developed in the 2nd half of the 20th century for direct potentiometric measurements.

Principle

The basic principle of potentiometry is to determine the electrical potential difference between WE and RE. The difference of electrical potential between WE and RE is proportional to that of the ion activity logarithm, as shown by the Nernst-Donnan equation:

$$E_{cell} = E^0_{cell} - \frac{RT}{nF} \ln Q$$

where $\mathbf{E_{cell}}$ (V) represents the equilibrium cell potential, $\mathbf{E^0_{cell}}$, **R** presents universal gas constant, **T** presents absolute temperature (K), **n** presents the total no. of electrons taking part in the redox reaction, **F** represents the Faraday constant, and **Q** represents the law of mass action law of redox process.

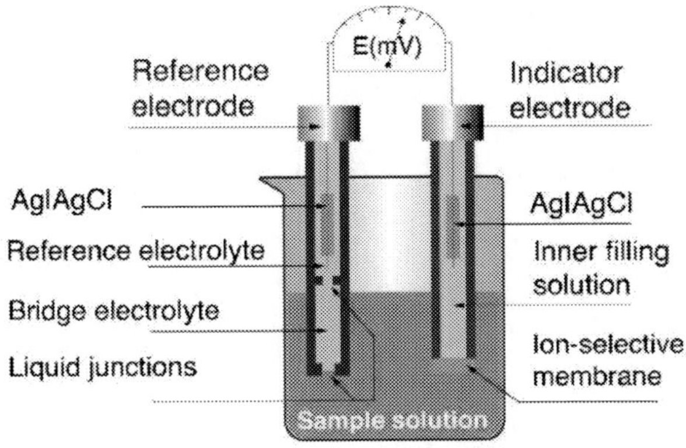

Figure 5.14. Schematic presentation of potentiometry.

Typical Arrangement for a Potentiometric Sensor

The metal electrode is in direct contact with the electrolyte solution in the 1st type of metal electrode. If these metal ions are present in the system, balance is achieved on the surface of the metal depending on the concentration in the solution of the metal ions. These metal ions were received by metal surfaces and discharged into the electrolytes at the same time. However, this electrode type suffers from poor selectivity since it usually reacts to any cations to which it is easily reduced.

A 2nd kind of metal electrode comprised of metal either covered or dipped in one of its salts, such as AgCl, generally responds to the salt anions.

A 3rd type of electrode only reacts to a cation other than that of a metal electrode. If the electrode has no ions of the relevant metals, then the oxidation/reduction potential for metallic electrodes can still occur when the redox reaction happens.

When the electrode surface is redox-inert, there are no metal ions discharged from the metal and for electrons, the metal surface acts as a catalyst. For metallic electrode indicator, platinum or gold is mostly used since it is chemically inert. This is the 4th kind of metallic electrode.

Currently, potentiometry utilizes electrodes that are sensitive to ions of interest selectively, for example fluoride selective electrodes. A filling solution is enclosed inside the membrane electrodes; this solution includes ions that the membrane is selective to. If there is change in the activity of these ions on both sides of the membrane, then ions come into the membrane from the side that has higher activity and come out of the opposite side of the membrane. This ion movement alters the electrical properties of the membrane and produces potential change. It is virtually never feasible to perfectly pick one ion type. Most ISEs react with ions having similar chemical characteristics or structures. The membrane of these electrodes is either made of an aqueous solution of Ca^{2+} (liquid material) or crystalline inorganic salts, such as LaF_3, Ag_2S, or non-crystalline such as polymer, glass (crystalline and non-crystalline are solid materials). The working ranges are from 0.1-1 M to 10-5–10-11 M for ISEs, which is much greater than that of other analytical techniques.

Limitation

Accessing the activity of analytes using potentiometry has many drawbacks:

1. The standard state potential (E^o) is dependent on temperature, whereas the values in reference table are according to the 25 °C temperature.
2. E^o is also dependent on the chemical composition of the solution.
3. Cell may be supplied with a small amount of current.
4. At the interface between salt bridge and sample, there is the development of junction potentials (i.e., at the interface of two ionic solutions).
5. Because of all these limitations, 1 to 10% accuracies are observed, which depends on the type of ion specie. Its precision is lower than ±0.8%, which is also dependent on the type of analyte.

Applications

1. For pH-measuring of environmental water samples, potentiometric methods are frequently utilized.
2. Potentiometric techniques are commonly used in clinical applications to access the activity of analyte, like Ca^{2+}, K^+, Na^+ and Cl^-.
3. Gas sensing is another important use of potentiometric sensors.
4. For the estimation of glucose, urea and uric acid, the potentiometric biosensors are more versatile.

Types of Working Electrodes Used in the Electrochemical Analysis (Carbon Nanocomposite-Based Electrodes)

Working electrodes (WEs): It is the main part of electrochemical cell, chemical and electrical properties of WE plays an important role to carry out the electrochemical reaction in electrochemical devices such as electrochemical sensors and electrochemical biosensors. The electrochemical processes occur on the electrode surface, so that the performance of the electrode can be enhanced by using the method of either surface activation or surface modification.

Surface activation of electrodes either through chemical or electrochemical:

1. Can remove the contaminants from the surface of electrode.
2. Can increase the surface area.
3. Can increase the functional groups on the electrode surface.
4. Enhance the behavior of electrode by exposing a fresh surface.

In the case of electrode modification, nanotechnology has been making a revolution in the last decades. Different types of working electrodes (WE) can be used in the electroanalysis procedure.

Mercury Capillary Electrodes

During the review of electrochemical methods, it was determined that owing to favorable benefits such as renewable surface, amalgam production, wide range of cathodic potential and greater potential to reduce the proton, notably polarography, was combined with mercury electrode.

The dropping mercury electrode (DME) is a working electrode mostly comprised of a mercury drop that is suspended from the end of the capillary tube due to gravity that provides a fresh surface to the electrode.

A desirable size of mercury drop is achieved by passing through a narrow tube via micrometer screw, in the hanging mercury drop electrode (HMDE).

The mercury film electrode (MFE) is constructed of a thin coating of mercury on the solid platinum or carbon electrode surface. The tri-mode electrode (DME or HMDE) is used in static mercury drop electrode (SMDE), and a solenoid-driven plunger is utilized to regulate the flow of mercury (Figure 5.15).

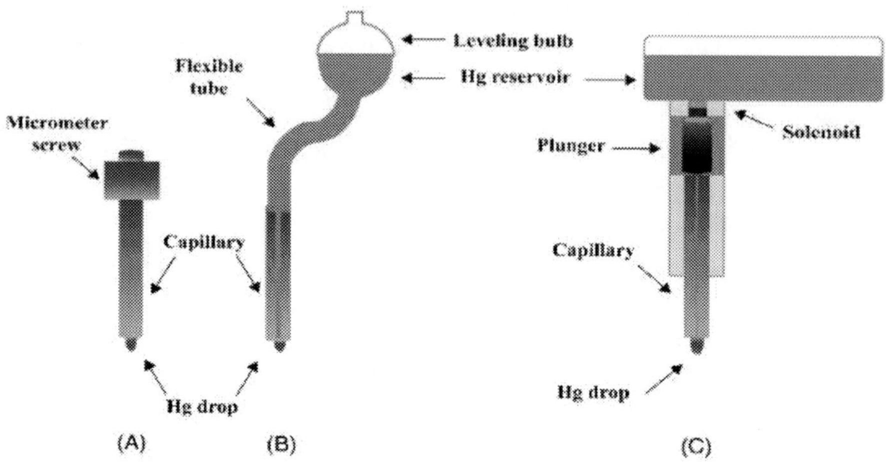

Figure 5.15. A schematic representation of mercury electrodes: (A) HMDE, (B) DME, and (C) SMDE.

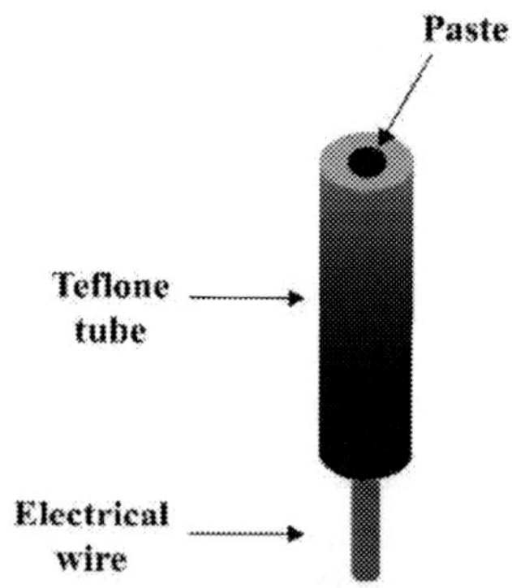

Figure 5.16. Carbon paste electrode.

Carbon Paste Electrodes (SPEs)

This type of electrode has valuable applications in electroanalytical procedures. CPEs are comprised of a cavity that is filled with carbon paste (carbon particles and viscous organic solution used for binding). The advantages of such electrodes include their ability to combine with various compounds, easy to renovate it and less waste of current. CPEs modified with CNMs, to enhance their electroanalytical performance and to reduce the overvoltage effect (Figure 5.16).

Glassy Carbon Electrodes (GCEs)

GCEs have distinguished electrical and mechanical properties, reproducible performance and chemical inertness. These electrodes are manufactured in an inert environment that is free from gases such as N_2, H_2 and O_2 with controlled heating program ranged from 300 to 1200°C. The usage of glassy carbon as an electrode in electrochemistry is broad, GCE consists of thin, tangle bands of high-density graphite sheets. The surface of GCE is continuously polished with 0.05 μm particles of either diamond or alumina in order to activate it and to increase its electroanalytical performance. Numerous treatments such as electrochemical, thermal, oxygen and chemical are used for further activation of GCE, which is required for removal of contaminants and functionalization of the surface.

Carbon Ceramic Electrodes (CCEs)

The CCE is a composite electrode, manufactured by attaching a silica matrix with powdered carbon. These CCEs offer several benefits, including their porous and rigid structure, easy to modify and they have high conductivity. In the electrochemical method, CCEs can be employed to identify or determine different compounds. New features can be obtained by the modification of carbon ceramic electrodes, like it can enhance the sensitivity of electrochemical sensors and improve the electroanalytical performance.

Graphite Rod or Pencil Graphite Electrode (GREs)

The GRE is another highly accepted and inexpensive carbon-based electrode, which has been successfully utilized to produce DNA hybridization biosensors to determine pharmaceuticals in samples, and used in other electrochemical devices such as in barriers.

Metallic Electrodes (MEs)

MEs are used over a wide range of voltages and are made up of platinum, copper, silver or gold. For example, in basic solutions, platinum electrodes have a possible range of potential from +0.7 to −1 versus SCE, and in acidic solutions, platinum electrodes haves a possible range of potential from +1.2 to −0.2 V versus SCE. MEs have a number of problems, electrode surface alteration due to electrolyte species absorption and deposition of oxide layer. MEs must be reconditioned through physical, chemical or by electrochemical polishing, in order to overcome these difficulties (Figure 5.17).

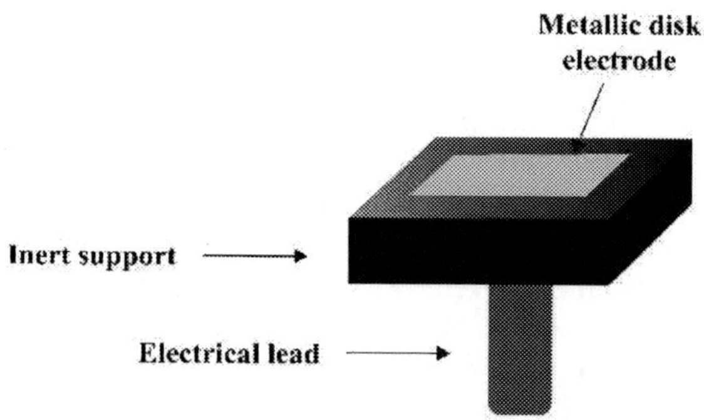

Figure 5.17. A schematic representation of metallic disk electrode.

Screen-Printed Electrode

This electrode is a chemical inert substrate, on which with the screen-printing three or more electrodes, i.e., pseudo-reference electrode (PRE), working electrode (WE) and counter electrode (CE) are printed. Layer by layer deposition of different inks on the solid substrate was carried out throughout the screening-printing procedure. In order to analyze different compounds in-situ, especially in the case of pharmaceutical samples, screening-printing techniques are important. SPEs (Figure 5.18) generally have dimensions of 3×1×0.05 cm, although they may vary depending on the material utilized, electrodes applications and insulating material. These electrodes have key benefits, such as appropriate rates of mass transfer, reproducibility, uniformity and stability. Different factors must be considered to produce an applicable SPE, including its size and dimension, which can be regulated by substrate material, screen, ink type, printing and drying. Glass, plastic and polymeric material are used in SPEs as an insulating substrate. The most frequent materials utilized to amend the WE in screen-printed electrodes, are platinum ink, gold ink, carbon ink and other form of carbons, i.e., graphene, graphite, fullerene and CNTs. For the detection of electrochemical compounds, the surface of WE is altered by using different materials. Ag/AgCl is stable and easy to modify, so it is used as PRE. ACE comprises the same material as a WE, but external platinum wire or rod is used. In order to increase its electroanalytical performance and selectivity, the surface of SPEs can be modified by using different materials like biomolecules, polymeric materials or Schiff-bases.

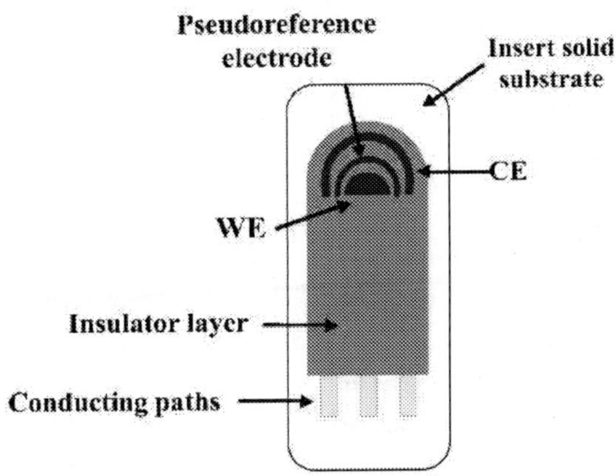

Figure 5.18. A schematic representation of SPE.

In general, two types of modifications of screen-printing electrodes are possible. First, with the use of different substances, variation in composition of the inks, and second, by modifying the prepared SPEs through depositing preferable substances on the surface of the electrodes which can enhance the electrode's performance by increasing its conductivity and selectivity.

Transparent Conducting Electrodes

Thin film of TCO (transparent conducting oxide) comprised of either binary compound or ternary compound like In_2O_3, Cd_2SnO_4, SnO_2, SnO_2:Sn and SnO_2:F are utilized in solar cells and spectro-electrochemistry. Transparent materials of different kinds including polyethylene terephthalate (conducting polymer), quartz and glass, are used to support these films. By CNMs TCEs are modified, this strategy significantly increases its conductivity as well as electroanalytical properties (Figure 5.19).

Figure 5.19. A schematic representation of TCO and GO/FTO.

Rotating Disk Electrode (RED)

Movement of electrodes with respect to the solution is necessary in several electrochemical methods. In these cases, either vibrating electrodes or rotating disk are used, since these are itself in motion or solution forcedly pass through a stationary electrode. RED is made up of platinum, or glass carbon, which is embedded into the insulating material such as plastic, epoxy resin, Teflon or glass. A motor is attached to this electrode, after this it rotates at frequency, *e* = $\omega/2\pi$, where ω presents angular velocity. In hydrodynamic voltammetry or hydrodynamic amperometry, REDs are most commonly used (Figure 5.20).

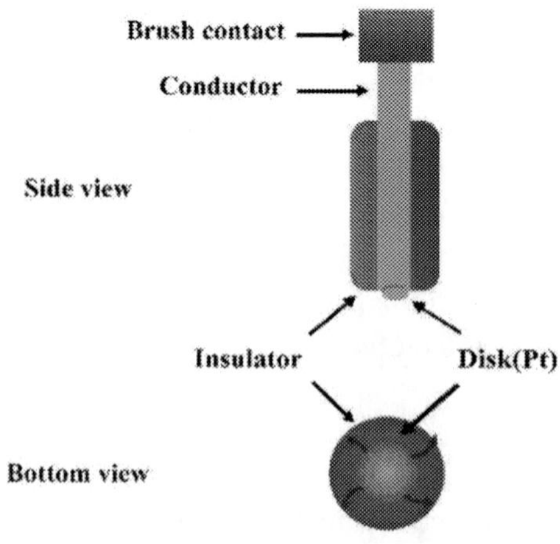

Figure 5.20. A schematic representation of RDE.

Rotating Ring-Disk Electrode (RRDEs)

For reverse methods, RDEs cannot be applied since the electrochemical reaction product is constantly swept off the disk, thus a concentric ring electrode can be placed to encircle the disk, which creates RRDE. Because in RRDE experiments, basically two voltages ED (voltage of disk) and ER (voltage of ring), and two currents ID (current of disk) and IR (current of ring) are monitored. Furthermore, RRE (rotating ring electrode) are used alone as electrodes (Figure 5.21).

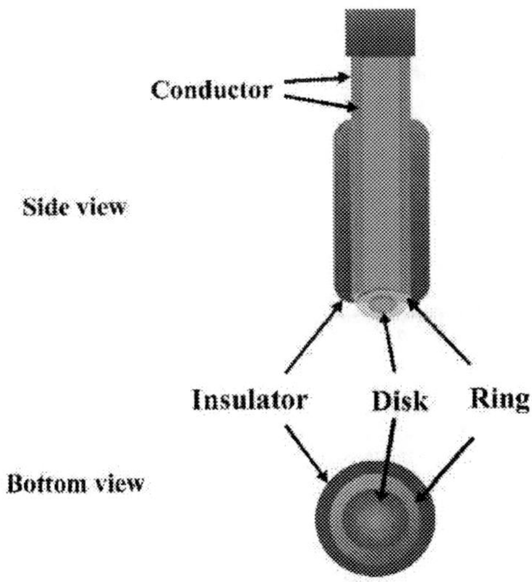

Figure 5.21. A schematic representation of RRDE.

Conclusion

Electrolysis is the 2nd most common type of instrumental analysis. Electrochemical analytical procedures make use of electrodes, which are electrically conducting probes that establish electrical contact with the solution in which analyte is present. The electrodes are joined with the electronic device to measure an electrical current and other solution parameters. We identify the analyte and estimate the amount of analyte in solution with the help of measured parameters. On the basis of measured electrical parameters, electrochemical techniques are classified as electrogravimetry, coulometry, polarography, voltammetry, conductometry, amperometry or potentiometry. They are all discussed in this chapter with their fundamental principles, the instrumentation, electrode types, limitation, advantages and application. Electrogravimetry, is the electrochemical technique, in which deposition of weighed material on one of the electrodes. In coulometry, the amount of current is measured and used by the analyte during an electrochemical reaction. Conductometry measures conductance to monitor a progress of chemical reaction, while between electrodes maintaining a constant AC potential. In voltammetry, current is monitored, whereas potential is varied in regular manner. Polarography is the subtype of voltammetry, in which a liquid metallic electrode is used. In amperometry, the current (amperes) is measured while maintaining a constant potential between electrodes. Potentiometry monitors voltage while keeping the electric current constant between electrodes.

The flow of electrons between one electrode or many electrodes and the analyte is the basis of most electrochemical reactions. Analytes must have the ability to be reduced (gain one or more electrons from electrode) or oxidized (donate one or more electrons to the electrode). After explaining all techniques, at the end of chapter different important types of electrodes have been reviewed.

Conflict of Interest

None to Declare

Consent for Publication

None to Declare

Acknowledgement

None to Declare

References

Bond, A.M., 2.15 - Electrochemistry: General Introduction, in *Comprehensive Coordination Chemistry* II, J.A. McCleverty and T.J. Meyer, Editors. 2003, Pergamon: Oxford. p. 197-222.

Bourkane, S., C. Gabrielli, and M. Keddam, Study of electrochemical phase formation and dissolution by ac quartz electrogravimetry. *Electrochimica Acta,* 1989. 34(8): p. 1081-1092.

Ciobanu, M., et al., 1 - Fundamentals, in *Handbook of Electrochemistry,* C.G. Zoski, Editor. 2007, Elsevier: Amsterdam. p. 3-29.

Frenzel, W., *Book Review: Principles and Practice of Analytical Chemistry.* 5th Edition. By F. W. Fifield and D. Kealey. Angewandte Chemie-international Edition - Angew Chem Int ed, 2002. 41: p. 1249-1250.

Gabrielli, C., et al., Ions and solvent transport across conducting polymers investigated by ac electrogravimetry. Application to polyaniline. *Journal of Electroanalytical Chemistry,* 2000. 485(2): p. 101-113.

Hauser, P.C., Coulometry, in *Encyclopedia of Analytical Science* (Second Edition), P. Worsfold, A. Townshend, and C. Poole, Editors. 2005, Elsevier: Oxford. p. 234-240.

Hillman, A.R., Electrogravimetry, in *Encyclopedia of Analytical Science* (Second Edition), P. Worsfold, A. Townshend, and C. Poole, Editors. 2005, Elsevier: Oxford. p. 317-324.

Hills, G.J. and D.J.G. Ives, The Hydrogen Electrode. *Nature,* 1949. 163(4156): p. 997-997.

Houssin, T. and V. Senez, Chapter Six - Electrochemical Detection, in *Waterborne Pathogens,* H. Bridle, Editor. 2014, Academic Press: Amsterdam. p. 147-188.

Hsueh, C.-J., et al., Coulometric Analysis, *in Encyclopedia of Applied Electrochemistry,* G. Kreysa, K.-i. Ota, and R.F. Savinell, Editors. 2014, Springer New York: New York, NY. p. 275-283.

Moldoveanu, S.C. and V. David, Chapter 2 - Short Overviews of Analytical Techniques Not Containing an Independent Separation Step, in *Selection of the HPLC Method in Chemical Analysis*, S.C. Moldoveanu and V. David, Editors. 2017, Elsevier: Boston. p. 31-53.

Pietrzyk, D.J. and C.W. Frank, Chapter Twenty-Eight - Introduction to Electrochemistry, in *Analytical Chemistry,* D.J. Pietrzyk and C.W. Frank, Editors. 1979, Academic Press. p. 581-597.

Sudha, P.D.C., *Pharmaceutical Analysis.* 2012: Pearson Education India.

Taherpour, A.A. and F. Mousavi, Chapter 6 - Carbon nanomaterials for electroanalysis in pharmaceutical applications, in *Fullerens, Graphenes and Nanotubes,* A.M. Grumezescu, Editor. 2018, William Andrew Publishing. p. 169-225.

Thorburn Burns, D. and F. Szabadváry, History of Analytical Science, in *Encyclopedia of Analytical Science* (Second Edition), P. Worsfold, A. Townshend, and C. Poole, Editors. 2005, Elsevier: Oxford. p. 267-277.

Trivinho-Strixino, F., J.S. Santos, and M. Souza Sikora, 3 - Electrochemical Synthesis of Nanostructured Materials, in *Nanostructures,* A.L. Da Róz, et al., Editors. 2017, William Andrew Publishing. p. 53-103.

Chapter 6

Radiochemical Methods of Analysis

Malik Saadullah[1,*], Jahan Ara Batool[1], Muhammad Rashad[1], Sameer Alshehri[2], Bashir Ahmad Chaudhary[3], Sheikh Abdur Rashid, Samina Afzal[3], Muhammad Ajmal Shah[4] and Bisma Kanwal[1]

[1]Department of Pharmaceutical Chemistry, Government College University, Faisalabad, Pakistan
[2]Department of Pharmaceutics and Industrial Pharmacy, College of Pharmacy, Taif University, Saudi Arabia
[3]Faculty of Pharmacy, Bahauddin Zakariya University, Multan, Pakistan
[4]Faculty of Pharmacy, Government College University, Faisalabad, Pakistan

Abstract

Radiochemical analysis automation has many advantages, including less exposure of workers to radioactivity, more consistent analytical protocols, better safety, and higher reliability. Before the quantification and detection of radionuclides, automation is significant in addressing criteria and requirements to carry out the radiochemical separations. These radiochemical analysis aspects can be time consuming and particularly expensive. Classical methods for the preparation of samples in radiochemistry require manual separations including extraction and precipitation. New radiochemical separations are carried out on materials of solid-phase separation, applicable for automation and have many advantages.

[*]Corresponding Author's Email: Maliksaadullah@gcuf.edu.pk.

In: Functional Analytical Techniques in Pharmaceutical Chemistry
Editor: Malik Saadullah
ISBN: 979-8-88697-059-3

Keywords: Activation analysis (AA), neutron activation analysis (NAA), spontaneous fission (SF), alpha (**α**), beta (**β**), gamma (**γ**), electron capture (EC), gas electron multiplier (GEM), microstrip gas chamber (MSGC), micropattern gas detector (MPGD) and Geiger Muller (GM)

Introduction

Historically, the radiochemistry field is introduced as chemical reaction studies using radioisotopes. Radiochemistry is based on the pillar of tracer principle and its methodologies are always originated from it. These range from geological, chemical, analytical, environmental and biological applications.

What Is Radiochemistry?

Radiochemistry is the branch of chemistry, which studies the chemistry of radioactive substances and their transformations, such as transuranium and actinide elements, solving the problems of radioecology, developing the physicochemical principles to handle radioactive waste from nuclear power engineering, separating the radioactive isotopes and developing the methods for the manufacturing sources of radioactive emissions, which is fairly significant.

What Is Radiochemical Analysis?

The analysis of radiochemistry is based upon two outstanding radioactivity features or characteristics. The radioactive radiations that are highly sensitive and easily measured, and the possibility of the labelled chemical compounds along with the radioactive tracers.

Radiochemical methods are used to study or investigate radioactivity in materials and compounds of radionuclides that are produced by irradiation and those radioactive materials that are naturally occurring. The foundation or establishment of studies of radioactivity is carefully measured in a wide variety of radiopharmaceutical, food and environmental samples, etc.

Such measurements may be divided into two major types:

Naturally occurring radionuclides measurement which include the objectives; determine and identify the amount of natural radionuclides, and by the use of that information, determine the amount and presence of the sample's related elements e.g., uranium, potassium and thorium.

Man-made measurements mean the artificial radionuclides that are involved to determine and identify the amounts e.g., fission products containing the radioactive spills, fallout and leakage from the sites of waste disposal or any other facilities.

General Principle of Radiochemical Analysis

Ionizing radiations are emitted in γ-ray spectrometry, nuclear particle and radioactive decay. Measurements of spectrometry and intensity by the qualitative and quantitative analysis, respectively.

Drawbacks of Radiochemical Analysis

1. Irradiation is frequently of a high cost.
2. It mostly involves the open use of radioactive sources which implies a radiochemical laboratory.
3. Many radioanalytical procedures require a long turn-over time.
4. Manual labour is required in most of these radioanalytical procedures.

Selection of Radiochemical Analysis

Selection of method is based upon:

1. Specific experiment
2. Type of radioisotope
3. Sample volume

Types of Radiochemical Analysis

Radiochemical analysis is of the following types.

Neutron Activation Analysis

Analysis of neutron activation is the most common form, activation of proton and charged particle analysis methods are applied. Activation analysis is a nondestructive and highly sensitive method for the determination of a sample's atomic composition qualitatively and quantitatively. To determine the elements in complex samples i.e., archeological objects, minerals, environmental and biological samples, activation analysis is widely used. NAA is a functional technique and it provides an alternative from simple to tedious, destructive and difficult technique. This technique also has some limitations, i.e., demand for a source of strong irradiation. Simultaneous determination of complex samples and elements is the main advantage of NAA technique.

Principle of Neutron Activation Analysis

The basic principle of this technique is that the bombarded materials are analyzed with photons (gamma-rays), charged particles or neutrons and due to this bombardment from target material elements radionuclides are produced. After that, the radionuclides are analyzed by use of radioassay methods qualitatively and quantitatively. After the radiation, bombarded radioactive products are measured and the energies half-life and emitted types provide information regarding use for the qualitative analysis and radiation intensity gives data for material having quantitative composition to be analyzed.

Basically, sample is irradiated to produce R amount of radioactive nuclide stated by the following relationship:

$$R = \psi\varphi\sigma N\left(1 - e^{-\lambda tirr}\right)e^{-\lambda t_{cool}}$$

Here it is assumed that in a neutron reactor, homogeneous particle flux i.e., $\boldsymbol{\varphi}$ produces irradiation. A minimum element amount is detected and there is an increase in measuring apparatus efficiency $\boldsymbol{\psi}$, reaction cross-section $\boldsymbol{\sigma}$, bombarding flux $\boldsymbol{\varphi}$, decay constant *l*, of formed radioactive nuclide, irradiation time $\boldsymbol{t_{irr}}$ till saturation activity, and time from ending of bombardment until the start of the counting, $\boldsymbol{t_{cool}}$. Properly select $\boldsymbol{t_{irr}}$ and $\boldsymbol{t_{cool}}$ changes an element's sensitivity and minimizes interferences.

Applications of Neutron Activation Analysis (NAA)

With its vast variety of applications, NAA is applied to material types, including river water, pharmaceutical and petroleum products, urine, particulates of air pollution, food products, feces, sediments, plastics, soils, marine samples, vegetation, coal, semiconductor materials, clays, metals, alloys, glasses, ceramics, and detects the trace impurities. The main advantage of this technique is that dissolution of sample is not required.

Isotope Dilution Analysis

There are three types of isotope dilution methods. These are direct, inverse and double isotope dilution.

Radioactive Isotopes Dilution or Direct Dilution of Isotope

This method constitutes many quantitative applications in organic or inorganic analysis and this technique is based upon the variations in tracer's early specific activity because of dilution with an inactive compound which is under investigation.

(1) The addition of a lesser amount of an active compound is the simplest case which is compared with an unknown inactive substance, or it is disregarded. In such case, the recovered amount of an active compound is directly related to recovered inactive compound. Hence the calculation is done by following equation:

$$W_2 = W_3 \frac{A_1}{A_3}$$

$\boldsymbol{W_2}$ is a determined element amount; $\boldsymbol{W_3}$ is a recovered element amount; $\boldsymbol{A_1}$ is activity introduce with the tracer; $\boldsymbol{A_3}$ is the activity of a recovered compound.

(2) If an amount of diluting substance is disregarded, amount $\boldsymbol{W_1}$ of radioactive isotope is then added into an unknown amount $\boldsymbol{W_2}$of elements which is to be determined, and particular activity of $\boldsymbol{W_1}$, have value of $\boldsymbol{S_1} =$

$\frac{A_1}{W_1}$, is being determined previously. The following equation is used to calculate the amount of an element which is to be determined:

$$W_2 = W_1 \left(\frac{S_1}{S_2} - 1 \right)$$

$\boldsymbol{W_1}$ and $\boldsymbol{S_1}$ are known, and after determining $\mathbf{S_2}$, it is easy for the calculation of $\boldsymbol{W_2}$ by an equation.

The calculation of $S_2 = S_3 = A_3/W_3$ is generally prepared by precipitation, a specific part of the mixture $(W_1 + W33_2) = W_3$, then weigh and measure the precipitate and its activity, respectively.

Inactive Isotopes Dilution or Reverse Dilution of Isotopes

In isotopes' dilution method modification, the reverse dilution method of isotopes is used to determine the small amount of substances that are not recovered in their natural form from a complicated multicomponent. Inactive isotopes' dilution is used to determine the yield of radioactive isotopes on nuclear transformations involving radiochemical investigations. In radioactivation analysis, for recovery and determining the radioactive isotopes, inactive carriers are widely used.

(1) The first approach is to determine the compound's radioactivity along with the other active substances in a mixture. After that, an inactive isotope carrier is introduced into an investigated solution, or in the greatly exceeded amount to that of the active compound, and also determined element is recovered. In such case, the recovered carrier quantity or amount is the direct measurement of recovered active compound. The calculation is done by following equation:

$$A_1 = A_2 \frac{W_{add}}{W_{rec}}$$

$\boldsymbol{A_1}$ is the complete activity of determined element; $\boldsymbol{A_2}$ is the recovered precipitate activity; $\boldsymbol{W_{add}}$ is the carrier amount added; $\boldsymbol{W_{rec}}$ is the carrier amount recovered.

(2) The second approach for determining the known specific activity of separate radioactive compound with the other radioactive compounds present

in a mixture, analysis is carried out. In this analysis, specified inactive carrier's amount is added in a mixture, and mixture part for isotopes analysis is withdrawn and specific activity is determined. The initial substance amount is then calculated from equation 2 to make use of the following notations:

$\boldsymbol{W_2}$ determined the element amount; $\boldsymbol{S_1} = \frac{\boldsymbol{A_1}}{\boldsymbol{W_2}}$ for specific activity; $\boldsymbol{W_{add}}$ is the weight of added carrier; $\boldsymbol{S_2}$ is the specific activity upon dilution,

$$W_2 = \frac{S_2}{S_1}(W_{add} + W_2)$$

Then,

$$W_2 = W_{add}\left(\frac{S_2}{S_1 - S_2}\right)$$

If the added carrier amount present in a sample exceeds greatly, i.e., $\boldsymbol{W_{add}} \geq \boldsymbol{W_2}$, so $\boldsymbol{W_{add}}$, is introduced in an equation, which is mentioned below, instead of sum of $(\boldsymbol{W_{add}} + \boldsymbol{W_2})$, desired equation becomes:

$$W_2 = \frac{S_2}{S_1}(W_{add})$$

This alternative requires knowledge about early specific activity, and because of that reason it is used in the metabolic investigation, in which there is difficulty in direct quantitative recovery and that's why it's used to determine the small amount of substance. This method is also use to check the compound's stability with labelled radioisotopes and the clarity of radioactive tracers.

Inactive Isotopes of Multiple Dilution or Double Dilution of Isotope

Unknown early specific activity occurs in element amounts that are determined by the multiple isotopes dilution contained investigating mixture aliquots. A different analysis is done, but inactive isotopes with known quantities are introduced in two uniform investigating mixture aliquots i.e., ten or twenty-fold excess compared with recommended element amount. When a mixture containing parts i.e., $(\boldsymbol{W_2} + \boldsymbol{W_{add1}})$ and $(\boldsymbol{W_2} + \boldsymbol{W_{add2}})$ are excluded, specified activities $\boldsymbol{S_1}$ and $\boldsymbol{S_2}$ are easily measured. The formula for calculation is described as:

$$\frac{W_2 = S_2 W_{add_2} - S_2 W_{add_1}}{S_1 - S_2}$$

For the case where equal carrier amounts are introduced in radioactive substances containing different amounts, analogous equations are obtained. The working possibility with the micro-concentrations (< 1 mg) is a major advantage in this method which is compared to that described above.

Application of Isotope Dilution Method

Applications for isotope dilution procedure are:

1. *Biochemistry*, to analyze complex mixture substances, significantly important in a biological standpoint, that include insecticides, vitamins and antibiotics.
2. *Organic chemistry*, for complex mixture analysis, i.e., fatty acids, amino acids and polymers.
3. *Inorganic chemistry*, to examine conventional analytical methods, to determine quantitatively difficult separable elements that exhibits the similar properties, and to determine quantitatively non detectable elements (rare elements) by the provided methods.

Tracer Method

In radiochemical determinations, radioactive tracers are introduced in unknown samples for radiochemical determinations and after this, the samples' activity of relative counts number is determined in tracer peaks and the analyte. The tracer method determines the chemical recovery and is an important check of quality control. Additionally, the counting efficiency must be known in those cases where separate measurements are needed to determine tracer.

In the early 1900s, George de Hevesy first designed the radioactive tracer method. De Hevesy was employed at the lab of Ernest Rutherford stated by

Van Houten (2002). Moreover, Rutherford was curious to study the properties of a substance named radium-D. He obtained radium-D mixed with lead which was a major problem. Rutherford was unable to separate the radium-D, therefore he challenged De Hevesy to the task. After that, de Hevesy attempted radium-D separation unsuccessfully for more than one year before he got an idea that if a substance is not separated from lead, it can be used as tracer that follows lead through different chemical reactions by doing radioactivity measurements and it was the birth of a method of radioactive tracer that finally led to a noble prize for De Hevesy in 1943, due to his work on the isotopes used as tracers for the study of chemical processes. De Hevesy was the person who discovered radium-D, which is actually a lead isotope, namely ^{210}Pb.

For a long time the initial concept of De Hevesy's radioactive tracer methods have been used to measure a wide variety of corrosion, erosion and wear mechanisms by industry.

The chemical behavior of radioactive isotopes is close to those of stable isotopes and is a fundamental principle in the method of radioactive tracers in the chemical process. The difference in radioactive atomic mass and atomic mass of stable isotopes provides a minor change in both at equilibrium constant and at the rate of chemical reaction, and it is an isotope effect that forms the basis of isotope separation.

Principle of the Tracer Method

The basic principle involved is that a volume or surface of tested material is first labeled with radioactive isotopes. Isotopes, according to Ehmann in 1991, is defined as "a form of element with the same proton numbers, but a difference in their mass due to different neutron numbers present in their nuclei." An unstable form of elements are isotopes which emit emanations i.e., gamma rays, beta or alpha particles when they decay. These emanations are radioactive and are measured by using different types of detectors.

Applications of the Tracer Method

Radioactive tracer technique consists of high sensitivity and it is easily detected and determined. Conventional detectors easily detect the element amount in picograms(10^{-12} g) or less by an appropriate use of a carrier-free tracer. In fact, the different biomolecules or hormones in radioimmunoassay are determined in picogram. The samples' condition i.e., physical or chemical (chemical concentrations, pressure and temperature) does not affect the tracer's radiating properties.

This technique is used in biochemical, medical, environmental, geological and biological studies. Industrial process control and diagnosis in medicine are the major practical applications.

Radioactive Decay

Nuclides (X) are atomic nuclei present in specific isotopes and are characterized by positively charged protons i.e., Z, neutrons i.e., N and nucleus energy state. Nuclide is represented as mass (A) or atomic number (Z):

$$^{A}_{Z}X_{N}$$

A = Z + N is regarded as a distinct nuclide and has enough energy for lifetime measurement, more than 10^{-2} s. Isotope refers to a type of atom and nuclide referred as its nucleus. Same atomic number as nuclide, and different neutron numbers are called isotopes, hence atomic mass A:

$$^{A}_{Z}X_{N}, Z^{A+1}\, X_{N+1}$$

Radionuclides are nuclides consisting of spontaneous disintegrating properties. Radionuclides may be classified as stable or unstable nuclides, there are 1700 known nuclides, from which 1400 are radioactive and 300 are stable.

Radioactive Decay Law

Radioactive decay (first-order reaction) is defined as, the direct relationship between the number of decays in unit time and the presence of and number of nuclei. If decay rate is −dN/dt, then,

$$-dN/dt=\lambda N$$

λ is a decay constant, λ value is differ for every radionuclide species, N is the presence of radioactive nuclei number. It is easy to exhibit the disappearance rate of radioactive nuclei by the following equation:

$$N = N_0 e^{-\lambda t}$$

N is the presence of a radioactive nuclei number in time t, N_0 is the presence of nuclei number present when t = 0. Therefore, the radioactive decay process is defined as exponential law, and characteristic property (half-life) controls the nuclide activity i.e., half-life is a time period in which half of nuclide original atoms are disintegrated.

Types of Radioactive Decay

Radioactive decay types depend upon the particular nuclide involved. Characterization of radioactive decay by α-, β-, γ-radiation as described in Table 6.1.

Table 6.1. Type of radioactive decay with their mechanism of action

Type of Radioactive Decay	Mechanism of Action
Alpha decay	Alpha decay is basically the emission of helium nuclei.
Beta decay	Beta decay is the creation and emission of either positrons or electron, or the process of electron capture.
Gamma decay	Gamma decay is basically the emission of electromagnetic radiation where the transition occurs between energy levels of the same nucleus.
Internal conversion	Internal conversion, in which a nucleus loses its energy by the interaction of nuclear field with that of the orbital electrons, causing expulsion of an electron and ionization of the atom instead of γ-rays emission.
Spontaneous fission	Spontaneous fission is observed only in the heaviest nuclei, in which the nucleus dissociates spontaneously into two roughly equal parts. Spontaneous fission is accompanied by the emission of electromagnetic radiation and of neutrons.

Alpha Decay

In 1899, Ernest Rutherford was convinced of nuclei radioactive decay by alpha emission. An alpha particle is a helium nucleus with two neutrons with two protons that bind together in stable configuration. α particle on a nuclear scale is a massive particle relative to other radiations and each with 4 units mass, 2 unit positive charge, emitted primarily in heavy nuclei radioactive transformation i.e., uranium, thorium. α particle radiations possess greater kinetic energy with 4-8 million (MeV) electron volts, whose range are considered as monoenergetic. α particles are positively charged and have massive size, so they do not travel fast in some media. Distance in air covered by α radiation depends upon energy measure in micrometer in the case of more dense material i.e., human tissue, and typical values ranging from 2.5-7.5 cm.

Helium nucleus loss in a heavy 'overcrowded' nucleus is mostly accompanied by emitting γ-rays and is summarized as:

$$^{A}_{Z}X_{N} \rightarrow {}^{A-4}_{Z-2}Y_{N-2} + {}^{4}_{2}\alpha_{2}$$

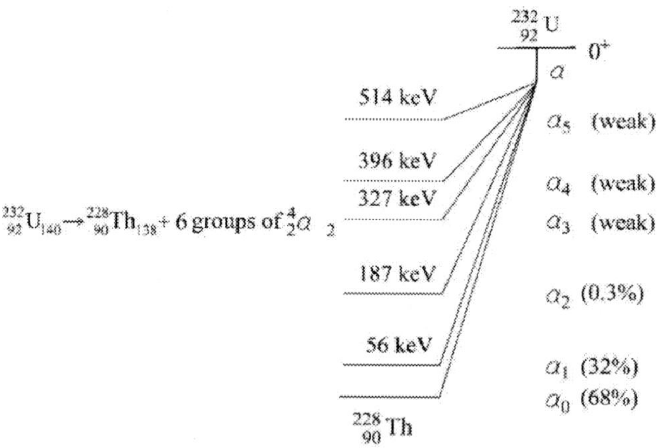

On the left, the equation exhibits uranium into thorium alpha decay and has six alpha particle groups decay spectrum that coincide with distinct energy level of α particles in U-nucleus. On the right, decay spectrum of α particles comprise intensity in brackets or kinetic energy as KeV.

Beta Decay

Weak interactions cause beta decay radioactive nuclides. Beta decay occurs when the ratio in neutrons and protons is non-optimal and thus allows the nucleus to reach an optimal ratio of proton/neutron. When there is excessive neutrons relate to protons, $(n \rightarrow p^{+} + \beta^{-} + \bar{v})$, than negative *β decay occurs and* when excessive protons relate to neutrons, $(p^{+} \rightarrow n + \beta^{+} + v)$, than positive *β decay occurs*. In beta decay, the atomic number is changed but the atomic mass number remains same. The atomic number decreases or increases in positive or negative *β decay* respectively.

In beta decay, rather than beta particle, the other particle is emitted i.e., neutrino in positive *β* decay or antineutrino in negative *β* decay.

Negative Beta Decay

$$^{A}_{Z}M \rightarrow {}^{\;\;A}_{Z+1}M + \beta^{-} + \overline{v}$$

Negative *β* decay process doesn't mean free nucleons, although bound in nucleus and neutron's rest mass is greater as compared to proton's rest mass, mass difference produces energy. **β⁻decay** represents a symbol of negative *β* decay and is exothermic.

Positive Beta Decay

$$^{A}_{Z}M \rightarrow {}^{\;\;A}_{Z-1}M + \beta^{+} + v$$

In positive *β* decay i.e., symbolized by **β^{+}-decay**, proton is transformed to a neutron. Positive *β* decay needs energy due to rest mass differences (1.3 MeV) and more energy (0.51 MeV) needed by positron for emission, both are provided by a decrease in nucleus mass, sum of two energies is 1.8 MeV and process of positive *β* decay is endothermic.

Gamma Decay

γ-rays emits from unstable nuclei that decay from levels of excited nuclei towards more stable ones. Energy emission as γ-radiations, unstable nucleus state falls to the more stable state. γ-radiation emitted by unstable nucleus atoms during radioactive decay are electromagnetic radiation and have many industrial applications due to higher energy that provides greater penetrability (Figure 6.1).

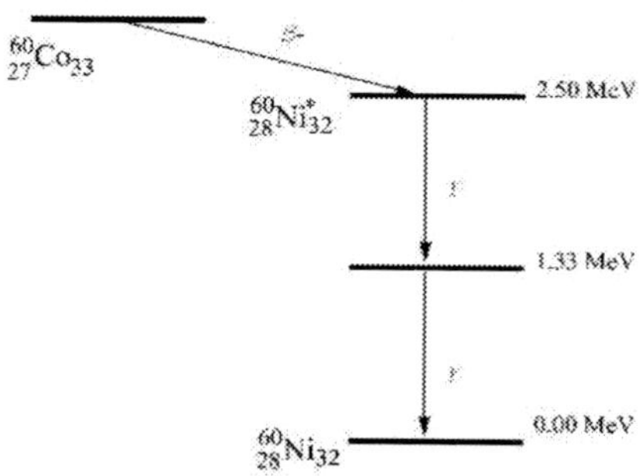

Figure 6.1. Schematic presentation of gamma decay of nickel with preceding formation of an excited state of nickel from cobalt.

Spontaneous Fission

Nuclear fission is a spontaneous fission (SF) induced by neutrons or photons captured by the nucleus. Decomposition of heavy nucleus occurs in nuclear fission (nuclear reaction) in two fractions named for several neutron or fission products.

Nuclear fission decay 100 nuclides by emitting neutrons as an alternative to another decay method, including *α-decay*. Spontaneously noninduced nucleus splitting into two nuclides is done by SF along with simultaneously emitting neutron more than average. This event mainly occurs with large mass number radionuclides i.e., $A \geq 230$.

Electron Capture

Electron capture, EC or EX decay, a process in which from inner electron shell i.e., K or L shell, nucleus captures electrons, resulting in transition:

$$\rho^{+} + e^{-} + \overline{\nu}$$

Electron capture decay is actively more advantageous than positive β decay, considering in EC decay no β particle emission. In electron capture process formation of neutrinos are monoenergetic. EC decay is followed by electromagnetic radiation emission in electron excited state due to orbital vacancy results. When a K shell vacancy is filled with outer electrons, i.e., L shell, the difference between emitted K or L binding energies are X-ray radiation characteristics.

Instrumentation of Radiochemical Methods of Analysis

1. Source (source of nuclear particles to induce radioactivity)
2. Cell (to hold sample)
3. Detector (monitor emission from sample by converting radioactive energy into electrical impulses)
4. Gas ionization detector (ion-chamber detector, proportional count detector, Geiger count detector)
5. Scintillation detector
6. Semisolid detector
7. Recording system (display pulse count)

Basically, the detector generates a number of electrical pulses with known time upon activation due to which the radioactive emissions are easily measured, i.e., count, is a measure of the sample's radioactivity. In some cases, the generated pulse height (size) is directly related with photons energy or the incident particles. To distinguish between different emissions and radionuclides, a pulse height analysis can thus be applied (Figure 6.2).

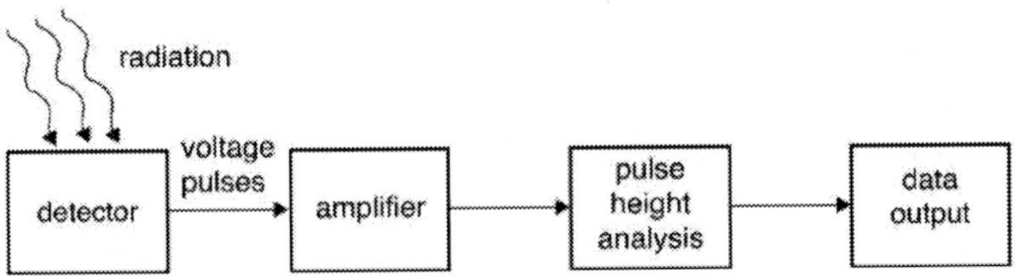

Figure 6.2. Schematic layout of radiation detection and measurement.

Detectors

Gas Ionization Detector

During the penetration of radiation into matter, radiation energy is passed above shielded radiation or may even be stopped. Atoms or molecules in matter show higher energy state which is excited state or high enough ionized radiation energy.

α − rays, β-rays, and γ-rays are ionization *radiations*. When pass through gas, create electrons or positive ions and these charged particles recombine then produces chemical reactions, and neutral specimens. In place of radiation, by applying electric field, positive ions migrate towards cathode, and so electrons are attracted to the anode. If voltage applied per unit of length and field strength is high to prevent the recombination during ions and electrons migration, all are arrived and collected at electrodes. Detection of electric charge indicates the presence of ionization radiation by utilizing suitable electric circuit.

Gas ionization detectors consist of gas volume in an enclosure which is constructed or sealed as it allows continuous flow of gas filling. An electric field is applied across electrodes, within that gas volume, the outer wall serves as electrodes, cathode, while in anode; wire, rod, grid, or plate is involved.

There are useful and interesting developments made over a past decade or in a field *of gas electron multiplier* i.e., GEM, *microstrip gas chamber* i.e., MSGC includes position-sensitive gaseous detectors with good localizing properties. These are *micropattern gas detectors* (MPGDs), due to accurate position with accuracy. Suitability of detector systems of gas ionization for radiation of given energy or kind depends upon ionized filling gas pressure and composition; applied field strength; geometry (shape, size, of electrodes and detector volume); thickness and type of construction material surrounded the detector gas volume.

Defining the Operating Characterization of Detectors of Gas Ionization

Few operating parameters are necessary for characterization of capabilities in case of the ionization detectors: resolving time, efficiency and resolution of detector.

Counting Efficiency

Counting efficiency is defined as the number of photons or particles emitted by radiation source related to number of interactions recorded by counting system and is known as absolute efficiency.

$$\textbf{Counting efficiency} = \frac{\text{number of signal record by detector}}{\text{number of photons or particles emit by source}}$$

Absolute Efficiency

Particles are not registered that strike detector volume, so other efficiency types are used, called intrinsic efficiency.

$$\textbf{Absolute efficiency} = \frac{\text{number of signal record by detector}}{\text{number of photons, particles strike the detector}}$$

Charged particles of absolute efficiency is 100% for ionization detectors. X-rays or γ-rays efficiency is reduced due to poor radiation interaction with gaseous detection volume that contains the ionization counters.

Resolving Time

It is defined as less time interval for a detector required to recover from a radiation interaction event or to register a following event. Resolving time in electronic system other than detector in counting devices e.g., data handling, steps of processing adjust limits for high-count rates dealing.

Energy Resolution

It characterizes the detector's discrimination ability between two radiation energy types that are distinct but close enough to each other. A characteristic figure given by full width at the half-maximum (FWHM), peak width displays halfway in radiation energy spectrum between peak top and its baseline. $\boldsymbol{E_0}$ is the energy at peak maximum, $\boldsymbol{\Delta E}$ is the full width at the half-maximum, hence resolution is described as $R = \Delta E/E_0$, that is recorded as % age. FWHM value is measure of resolution.

Localization

Some designs of the detector provide information regarding photons or particles entrance region or radioactive material distribution in the detector or in the sample respectively, which is significant. They provide radioactive specimen images by exhibiting radioactivity that is longitudinal or has 2D distribution.

Types of Gas Ionization Detector

Gas-filled detectors consist of three different types: Geiger-Muller counters, ion chambers and proportional counters. These detectors differ due to the application of electric field strength between the electrodes.

Ionization Chamber/Ion-Chamber

Ionization chambers are considered as the simplest devices to measure radioactivity. Ion-chambers used in early research days dealt with ionizing radiation detection. But today, new ionization design chambers have evolved for a special purpose. An ionization chamber is made of nonporous material, the electrodes consist of parallel plates, gas filling pressure ranges from a few tenths up to bars of some tens. Ion pairs are created when the radiation of ionization is passed through a gas. A sufficiently high-voltage gradient can be applied for the prevention of recombination, and these ions are drift towards electrodes. The recorded output signal by an electric circuit is a collected amount of total electrical charge, voltage pulse or charge and a current flow.

Ion Chamber Types

Ion chambers are mainly of three types; ion chambers operate in pulse and current mode and electrostatic ion chambers or charge integration.

When ion chambers operate in current mode, electric current flow is recorded, initiated by ions or electrons collected at electrodes during an observation time. While operating a pulse-mode chamber, individual signals including pulse voltage produced by ions reaching towards electrodes from a single event are registered or recorded by the application of circuits of suitable electronic amplification. Ion chambers of charge integration or electrostatic types are similar to electroscopes. A static electric charge is provided to a system that contains fiber or thin foil suspended in a parallel way to a solid support or second foil. Due to repulsive charges (negative-negative or positive-positive), foil or fiber will bend to stay some distance far from the second foil or fiber support. After that, radiation of ionization is gradually discharged from the system, which causes fibers or foils to move back into their original position.

Ion chambers have multiple applications due to their low cost and simple construction. Information can be obtained that is related by radiation type and energy, hence ion chambers are designed for detection of high and low radioactivity levels.

Proportional Counting of Detector of Gas Ionization

Proportional gas ionization detectors operate at a high-voltage gradient rather than ion chambers. Ion pairs (primary) are produced and accelerated by radiation of ionization and acquire high kinetic energy. Secondary ionization is induced by the collision of primary ions along with the drift of molecules or atoms. After secondary ionization induction, more energy, ions and electrons are released for further filling of the gas ionizing process. The multiplication process is referred to as the *Townsend cascade or Townsend avalanche.* The anode wire has to be thin enough to obtain sufficient field strength region. In a high-voltage gradient, gas amplification takes place near the anode wire. Amplification remains linearly proportional to the original ionization i.e., number of ions after gas amplification is directly related with primary ion number produced by ionizing radiation.

Proportional counting tubes are sealed with radiation sources remaining outside the tubes, a thin window allows radiation penetration to the volume detector. Another configuration is designed for gas and sample flow inserted into the volume detector so "windowless" counting systems can be used to detect α particles or less-energy β *particles*. Proportional counters operate in pulse mode and special gases or gaseous mixture are employed for proportional counters, gas filling does not produce anions and does not contain components that attract electrons, secondary Townsend avalanches formation is avoided. All noble gases optimally fulfill this requirement.

Applications of Proportional Counting Gas Ionization Detector

1. Proportional counters are mainly useful to discriminate between the α and β particles present in a mixed source.

2. Proportional counters of low level are widely used for the quantitative measurement of radionuclide after the procedures of radiochemical.
3. Produce output that is proportional to radiation energy.
4. Measure energy of incident radiation.

Geiger-Muller Detector

Geiger-Muller (GM) tubes applied higher electric fields than proportional counters and ion chambers. Gas filled GM tubes meet requirements equal to gas filled proportional counters. Argon and helium are mostly used. Gas pressure order is tenths of bars, due to depending upon tube size and shape; voltage (hundreds of volts) is applied in GM tubes. GM tubes operate at a pressure of low gas and are mostly sealed permanently. Information regarding the radiation energy and its types are only given when the shielding effect is observed related to radiation. α particles stop the thin layer matter, β particles exhibit the highest range for penetration into the shielding material. Before the entrance of β particles in a detector, photons exhibit a logarithmic decrease in intensity with the increase of material thickness.

Applications of Geiger-Muller Detector

Currently, GM tubes are used for contamination control in radiochemistry day-to-day work and radiation monitoring.

Photon doses (neutrons/gamma) in mixed fields can measure GM counters. But its response depends upon photon energy above 3 MeV.

Scintillation Detector

Scintillation counters constitute a more important group that depends upon the radiation absorption by the scintillator that makes UV light scintillations; first detect and then convert by photomultiplier to amplified pulse voltage (Figure 6.3).

There are two types of scintillation detectors. One is solid scintillators and other is liquid scintillators.

Solid scintillators are mainly used for analysis and detection of γ-rays and X-rays. The solid scintillators include the following materials; lithium and sodium iodide, naphthalene, anthracene and the 'loaded' polymers. Crystals of NaCl are activated by 1% incorporation of TlI that have been the cornerstone for many years of γ-ray spectrometry. When striking, crystal γ-rays interacts in many ways, most important are the Compton scattering and photoelectric absorption. In the former, fully absorbed γ-photons, and electrons give excitation to energies which directly relate with photon energy. Eventually, light scintillation intensity, along with voltage pulse height is produced, all are proportional to an incident photon energy.

Liquid scintillators found extensive employment of pure negatron emission measurement i.e., 14C, 3H, 32P, 35S, especially for low particle energy (<1 MeV), negatron carry small energy, i.e., 3H (0.018 MeV) and 14C (0.16 MeV).

To ensure ideal contact between scintillator and radionuclide, sample and scintillator are dissolved in the same solvent. Suitable solvents are xylene and toluene and retain scintillators in a solution in water for up to 20%. Insoluble samples are stabilized with the silica gel and are measured in suspensions. Solvent absorbs the energy of particles and transfers it to the scintillator, a photomultiplier is activated by re-emitted UV radiation pulse. A typical 'cocktail' constitutes secondary scintillator for liquid scintillation counting, shifting scintillations towards the larger wavelengths, in that manner, also shifts emission peaks for a photomultiplier to region of more sensitivity. Typical scintillators are considered as aromatic conjugated molecules. Scintillation intensity relates to incident ionized particle energy and hence an analysis of pulse height is employed to provide measurement of a spectrometric discrimination. Moreover, it is achieved or accomplished if there are large negatron energy differences (ten-fold). Significant interference are associated with liquid scintillation counting which are chemical or color quenching, chemiluminescence and bioluminescence.

Application of Scintillation Detector

1. *Solid scintillation:* Useful for γ-emitting isotopes.
2. *Liquid Scintillation*: For quantifying weak β- emitters.

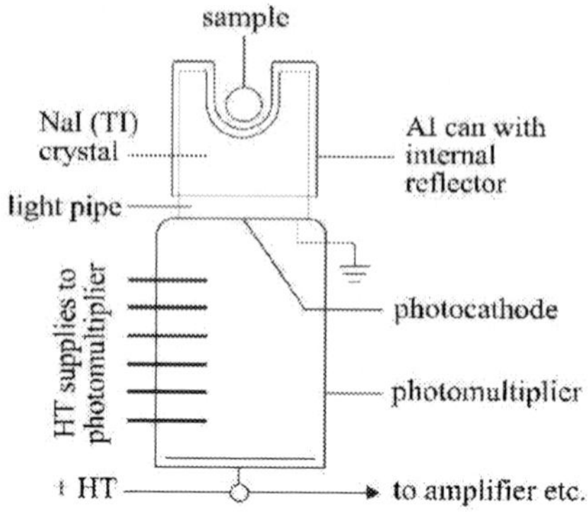

Figure 6.3. A typical scintillation detector. (Well crystal NaI).

Semiconductor Detectors

These detectors depend upon high-energy radiation interaction with semiconductor crystals like germanium or silicon. The former finds gamma radiation high or medium-energy and the latter for the X and low-energy gamma radiation. When crystals absorb radiation, production of positive holes and free electrons occurs and are directly proportional to incident photon energies. A gold thin layer is plated on electrodes and crystals of each end and 3-5 keV potential is applied across the crystal. Hence, positive holes and electrons produces electric pulses that are migrated towards electrodes and their sizes are directly proportional to incident energy radiation. Detectors exhibit significant importance in the X-ray spectrometry and its principle involves modern gamma-rays spectrometry.

Semiconductor detectors are p-n-junction devices produced to create two zones within semiconductor material. Material zone of *p-type* acts as electron acceptor from the valence band and material zone of *n-type* acts as electron donor to the conduction band. If electrical potential applied to detector that side p is negative rather than n side which is reverse bias, noise-generated current is flow, until, in detector, formation of electron-hole pairs by radiation interaction. This operating mode is used by semiconductor detectors (compensated or uncompensated).

Reverse bias applied in uncompensated detector to p-n junction and electric field sweep charges i.e., electrons and holes towards n and p side respectively from area between p- and n-material types, producing an electrically neutral insulate depleting region. No current flow in stable situation. Charges generate in depletion region from radiation event are collected and the current generate is proportional to energy interaction. Depletion depth increases and decreases by square root of applied bias and impurity concentration respectively. Ge or Si material purity determines depletion depth thickness. Commonly uncompensated p-n-junction detectors with depletion depths range from microns up-to few mm, made from silicon. Different semiconductor detector types are explained in Figure 6.4.

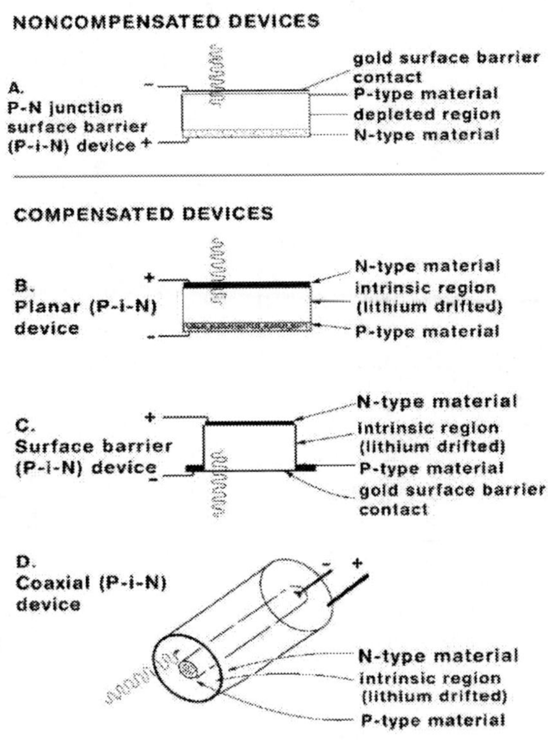

Figure 6.4. Various semiconductor detectors.

Types of Semiconductor Detectors

Semiconductor detectors are effective in different ways than the gas-filled counters and their density is 1000 times more than gas filling counters, stopping power for gamma or X-rays is also larger by same factor.

Semiconductor detectors have high resolution energy as its important feature, about 6-9 times better than gas counters and approximately 20 times better than NaI(TI) detectors. Different semiconductor detector types are illustrated in Table 6.2.

Table 6.2. Properties of Ge and Si semiconductor detectors and their applications

Type	Detector	Depletion thickness	Diameter (surface area)	Primary application in medicine
Uncompensated p-n-junction device	Si (p-n) (surface barrier)	a few microns to a few mm	1 to 50 mm (≈3 mm^2 to 2000mm^2)	Charged particle or low energy X-ray detector at room or body temperature.
	Ge (HP)	a few microns to 1 cm	up to ≈25 mm (≈500 mm^2)	X-ray or gamma ray detection when cooled to liquid-N_2 temperature. Can possibly be warmed up without damage to the detector.
Compensated p-n-junction device (p-i-n)	Ge (Li) (planer)	up to 1cm	up to 7 cm (≈39cm^2)	X-ray or gamma ray detection for energies above ≈20KeV. Must be maintained and operated at liquid-N_2 temperature.
	Ge (Li) (surface barrier)	up to 1cm	up 30 mm (700 mm^2)	Low energy x-ray or gamma ray detection. Must be maintained and operated at liquid-N_2 temperature.
	Ge (Li) (coaxial)	up to 7.5 cm	up to 6 cm (28cm^2)	High energy x-ray or gamma ray detection. Must be maintained and operated at liquid-N_2 temperature.

Type	Detector	Depletion thickness	Diameter (surface area)	Primary application in medicine
	Ge (Li) (surface barrier)	up to 5 mm	0.1 to 20 cm (≈0.8 mm^2 to 15 cm^2)	Charged particles and low energy x-ray or gamma ray detector (<40KeV) with thin entrance window.
	Ge (Li) (avalanche)	up to 200μ	1 to 16 mm (0.8 to 200 mm^2)	High sensitivity to charged particles and very soft x-rays. High count rate capacity (≈10^8counts/sec)

Application of Semiconductor Detectors

1. These detectors are used for charged-particle detection.
2. Application to analytical chemistry (radionuclides along with their analytical uses).
3. Different analytical problems are easily solved by the use of radioactive materials.
4. Isotope dilution analysis is done by using radiotracers.
5. It is difficult for the determination of specific component's exact amount present in compounds of complex mixtures (e.g., for the synthesis of biochemical or organic systems). A generally used technique is isotopic dilution analysis of considerable value, although, radiotracer technique is commonly used because of its simplicity and low-cost.
6. Small radio-chemically selected compound having a pure amount of reference ($\boldsymbol{w_0 g}$) is added in a compound of unknown amount ($\boldsymbol{w_u g}$) present in complex sample. The reference used in the above method is either a labelled compound or an element. (S_0 Bq /g) is known as the specific activity of reference substance. After complete mixing, the selected compound must be isolated in high purity rather than to obtain greater yield. A separate compound is to be weighed (w_m, g) or counted (R_m Bq) and specific activity (S_m
the following formula:

$$W_u = (S_0/S_m - 1)W_0$$

Activation Analysis

Activation analysis is known because of its high sensitivity and is a nondestructive technique mainly used to determine the atomic composition of sample qualitatively and quantitatively. Elements and the complex samples including archeological and biological objects, environmental samples and minerals can be easily determined through this method and it is a main advantage. This method also provides a simple alternate to more tedious, destructive and difficult techniques. A major drawback of activation analysis is the source of strong radiation. If sample counting analysis is being carried out with a multichannel analyzer or Ge-detector, more or dozens of elements are measured quantitatively. Decay properties in this technique like radioactive energy decay and half-life of a specific nuclide serves to identify uniquely. Analysis is being accomplished by radioactivity formation through sample irradiation by charged particles and also neutrons. Neutron irradiation technique is commonly used; hence method is referred to as NAA or neutron activation analysis.

Applications to General Chemistry

Chemical Pathway Studies

Radioactive tracers played a significant role in biological and chemical reaction pathway studies. The basic principle involves radioactive decay of the atom that announces its position. Radioactive nuclide, e.g., ^{14}C (β^- t½ 5730 y), 32P (β^- t½ 14.282 d), ^{198}Au ($\beta^-\gamma$, t½ 2.6935 d), follow through distinct steps involved in chemical reactions, revealed the details of a process or metabolic reactions that are difficult to discover by the other technique.

The analysis method requires analyte separation from the interfering materials that are present within the sample matrix. Different procedures affect these separations and quantitative separation establishment is a singularly beneficial step in a development method. Controlling of experimental conditions and careful studies is necessary and there is a need to establish larger number of experiments.

Radiotracers are most suitable for such studies. Tracer's detection sensitivity means the addition of a small quantity following the chemical pathway for specific species. Additionally, it matters that what state of tracer (chemical or physical) is made for measurements on gases, solids, liquids. Basically, separation methods including solvent extraction, precipitation and chromatography are widely studied as radiotracers. In individual separation steps, specie distribution can be studied simply by the radioactivity measurement, and tracer as yield indicator serves for all procedures (Figure 6.5).

Reaction steps involved in carbohydrate photosynthesis from an atmospheric CO_2 in sunlight and chlorophyll presence is a good example of this technique. Overall photosynthesis process (including many steps) written as:

$$6\,CO_2 + 12\,H_2O \xrightarrow[\text{chlorophyll}]{\text{light}} C_6H_{12}O_6 + 6O_2 + 6H_2O$$

By using tracers i.e., ^{14}C, ^{32}P, partition chromatographic technique, Calvin with his co-workers recognize that it involved intermediate steps. In first step, plants kept in atmosphere that contain ^{14}C-labelled CO_2, irradiated with light. Behind many irradiation times interval, plants removed and sections digest to dissolve material. After this, a few drops of solution in substance is separated (e.g., metal ion or organic molecule) and placed a few cm from the end of the paper strip. After that, the paper strip is hung vertically or dipped in solution so the initial placement point of substance is near to strip (bottom) above solution level. The solution travels upward due to the capillary force and then is brought into contact at the starting point with adsorbed substances. After paper development, the strip is removed from the tank, then substance moves for certain distance from starting point on paper and travelled with a distance depend upon type of paper, chemical properties of substance and solution used. Then Rf value for every substance is calculated as:

$$R_f = \frac{\text{distance the substance of interest has travelled}}{\text{distance the liquid front has travelled}}$$

Studies of Surfaces and Reactions in Solids

Solids having surface properties are studied by the radioactive tracer, which involves dipping the specimen into a suitable solution containing radioactive tracer. Then after exposure time, remove them, carefully wash the surface and then finally measure the emitted radiation from them. An extremely rapid exchange occurs between the solution metal ions or atoms on metal surface. While the exchange is the function of surface nature, within a minute it involves several hundred deep layers of atoms. The penetration depth of adsorbed radioactive isotopes are obtained by the measurement of radioactivity. Atomic diffusion in solid matrix is also studied in this technique.

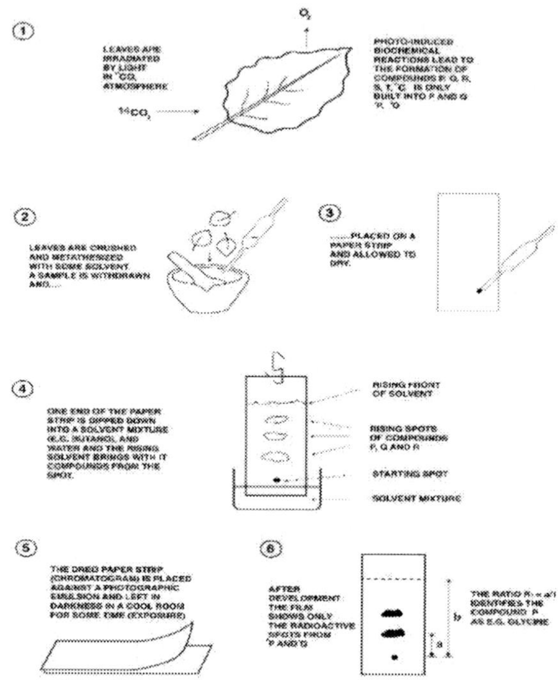

Figure 6.5. Paper chromatographic method for determination of which of components P, Q, R, S and T in a leaf are involved in photosynthesis.

Applications to Life Sciences

Various applications of radionuclide use in life sciences are highlighted in Figure 6.6.

A Radionuclide production

1 Target chemistry.
2 Irradiation
3 Isolation
4 and processing, yields
  (a) pure radionuclide, or
  (b) labelled compound (radiopharmaceuticals)

B Source, position/ administration

1 Internally (α, β or γ)
  (a) injection,
  (b) inhalation, or
  (c) oral intake.
2 Externally as -source

C Technique/application [detection]

1 Biochemical analysis
  (a) autoradiography [photographic]
  (b) immunoassay [counting]
  (c) DNA-analysis [photographic]
  (d) direct tracing [counting]
2 Medical imaging
  2.1 Transmission Tomography (TCT) photographic, or by counting computer display]
  2.2 Emission Computed Tomography (ECT) [counting computer display]
    (a) Single Photon Emission Computed Tomography (SPECT)
    (b) Positron Emission Tomography (PET)
3 Irradiation uses
  (a) by internal sources (therapeutically)
  (b) by external sources (therapeutically)
  (c) by external sources (sterilization etc)

Figure 6.6. Survey of radionuclide use in life sciences.

Radioimmunoassay

Radioactive markers are used in radioimmunoassay. In RIA, when radioactive markers are added, these radioactive markers compete with the sample analyte to bind with the antibody. Antibody or antigen molecules are labeled in case of radioimmunoassay with high-molecular weight of antigen. A radiolabeled second antibody is also common to use, i.e., antibody against primary antibody. In radioimmunoassay, labeled haptene is used for substances having low-molecular weight. Radioimmunoassay was developed in 1960s by using iodine isotopes, ^{125}I, ^{131}I, tritium (^{3}H) as labels. In clinical laboratories, RIAs use is declined due to concerns over safe handling and radioactive reagents or waste disposal. Compounds selection for radioimmunoassay are developed and explained in Table 6.3.

There are two different methods of RIA referred to as *RIA double-antibody* and a *coated-tube RIA.* Both are used in biological matrices for drug detection. RIA along with double-antibody, particularly second antibody, are introduced to facilitate the binding of primary antibody precipitation, if primary or secondary antibody–antigen complex precipitates, then the nonbounded label drug is easily removed. There is a coating of primary antibody in each tube side of coated-tube RIA, and then an unbound labelled drug is easily removed by supernatant pouring. In each RIA method, samples are analyzed by γ-counter for determination and investigation of per minute counts, and inversely related to original specimen's drug amount.

Table 6.3. A selection of compounds for which radioimmunoassays have been developed, indicating the wide applicability of the technique

Drugs	Steroids	Polypeptide hormones
Amphetamines Barbiturates Diazepam Digoxin Gentamicin Methadone Morphine Penicillin's Tubocurarine	Anabolics Androgens Corticosteroids Oestrogens Progesterones Steroid glucuronides	Adrenocorticotrophic hormone (ACTH) Follicle stimulating hormone (FSH) Growth hormone Luteinizing hormone (LH) Prolactin

First, it is required to obtain a specific binding agent to establish RIA method, which is done by the use of analyte to trigger the mechanism of natural defense of an animal containing antibodies produced, neutralizes foreign molecules like antigens or immunogens entering bloodstream. Antibodies are considered as specifically binding proteins i.e., g-globulin exhibited a small reactivity with other immunogens even if they have a close resemblance in structure. The analyte's relative molecular masses are 6000 or exceed producing the desired response, but small molecules i.e., haptene are needed to attached with suitable protein before injecting into an animal. Production of binding agent exhibited enough specificity for original analyte that provides the basis of RIA method. Once the level of antibody in the animal's blood is large enough for 6–12 weeks, serum, blood, and antibodies taken from animal are then separated. No more separation is required, nor is a binding agent used in that form. Serum produced antibodies are referred to as antiserum. Suitable antiserum formation is quite a difficult stage in order to develop RIA procedure, and requires a specialist with biochemical knowledge or skill.

Typical Procedures Involve the Following Stages

1. Solutions of standards or mixed samples with measured amount of antiserum and labelled analyte.

2. Incubation is followed to adjust the solution conditions e.g., at 2–4°C for about 1 1/2 hours for ensuring the reproducible and maximum reaction.
3. Free and bound analyte separation is done by free analyte adsorption on activated charcoal and protein binding fraction precipitation with ammonium sulphate.

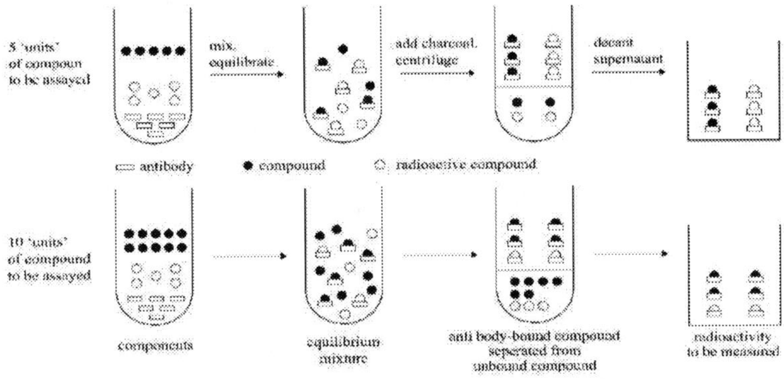

Figure 6.7. The principle of radioimmunoassay. In this example charcoal is used to adsorb the unbound compound and the amount of radioactive compound bound to antibody is measured.

4. Radioactivity ratio measurement in bound or free fractions. Activity ratio **A** relate with sample containing X analyte concentration.

$$A = \frac{Z}{(A + Y - Z)}$$

Y is the labelled amount and **Z** is the equivalent antiserum amount introduced which is bound. Simplicity, sensitivity and specificity of operation are considered to be the major advantages of RIA.

The principle of radioimmunoassay is explained in Figure 6.7.

Applications of RIA

To determine the organic compound present in complex matrices of small amounts is considered a tough analytical problem. If separation procedures are used, then these procedures are time consuming and complex. By the use of specific binding reagents, the method of RIA avoids difficulties and facilitates the compound analysis to picogram level. It is also used in imaging investigations (Figure 6.8).

Examination	Radionuclide	Static/ dynamic	Principal application
Bone	^{99m}Tc	S	Secondary spread of malignancy
Liver	^{99m}Tc	S	Secondary spread of malignancy; cirrhotic changes
Brain	^{99m}Tc	S/D	Occult metastases; brain damage; vascular problems
Lung	^{99m}Tc	S	Pulmonary embolism (perfusion)
Lung	^{133}Xe/^{99m}Tc	D	Pulmonary emphysema (ventilation)
Kidneys	^{123}I/^{99m}Tc	D	Renal function
Hepato-biliary	^{99m}Tc	D	Patency of the biliary tree; liver function system
Thyroid	^{123}I/^{99m}Tc	S	Thyroid function
Heart	^{201}Tl	S	Cardiac infarction and ischaemia (perfusion)
Heart	^{99m}Tc	D	Cardiac wall motion (blood pool)

Figure 6.8. Ten of the most frequently performed imaging investigations.

Environmental Applications

During the testing of nuclear weapons, large amounts of radionuclide have been released in the atmosphere, carried to Earth's surface by their own weight or by the rain. Geophysics use this weapon "fallout." To measure T (as HTO water), ^{90}Sr, ^{137}Cs, or other fission products and it is possible to follow water movements from land via rivers or lakes into sea, so it is also helpful for water stream oceans study and exchange between surface and deep water. Hence, Earth's water has been mapped in quite some detail.

Tracers involved in radioactivity i.e., T_2O, $^{24}Na+$, $^{82}Br^-$($\beta^-\gamma$ t½ 1.471 d), ^{51}Cr-EDTA($\beta^-\gamma$ t½ 27.704 d), are useful for the determination of natural water reserves volume in hydrology (underground) or also mapping movement of both surface or underground water along with effluent pathways. Due to this, flow and consumption of water in different industries is simply determined, and the leakage in pipes and dams is examined by radiotracers. A semiconductor detector containing analyzer of multichannel pulse height is used. Qualitative or quantitative analysis is done by peak energy or peak area.

Industrial Uses of Radiotracers

Industry has applied many radiotracers in large variety. In the United States, more than half of the 500 greatest production concerns use radioisotopes to produce different products like tobacco, plastics, food, metals, paper, clay, pharmaceuticals, chemicals, rubber, textiles and glass products. Radioisotopes study the chamber geometry effect, mixing efficiency, residence time in the reactors, towers for the fractionation, racemization, absorption, patterns and flow rates in columns.

General Applications

A patient's blood volume is determined by ^{32}P and is a general application. For that purpose, patient's 15 mL blood is withdrawn and mixed with $Na_2H_{32}PO_4$ small volume with high specific activity. RBCs (erythrocytes) in one hour take up all ^{32}P and 1 mL of activity 216000 cpm in detector system is generally used. An exactly tagged sample of 5 mL is then reinjected into a patient, and then 30 minutes later withdraw a new sample; hence 10 mL gives 2300 cpm and the volume of blood is calculated.

Amino acid mixture for cysteine is analyzed. 1 mL sample (density of 1 gmL^{-1}) is withdrawn, with ^{35}S 2.61 mg labelled cysteine of 0.862 μ Ci mg^{-1} specific activity is added. Isolate pure cysteine by liquid partition chromatography from this mixture, 30.6 mg is isolated that measures to provide 169000 cpm with efficiency of 27% detection. Then calculate cysteine percentage in original mixture.

By the use of radioactive tracer lead content having color pigment is mainly determined. To determine color pigment of lead content, 8.9871 g dissolved in conc. HNO_3 along with the addition of 5 mL ^{210}Pb solution. After the removal of excess acid throughout the process of evaporation, add 1M NaCl in excess, then filter the heated solution. After the cooling and crystallization process, washed $PbCl_2$ is then recrystallized, and the crystals of 0.3276 g is measured in 5 minutes in scintillation counter, giving 185160 counts. 1ml ^{210}Pb original solution gives 57000 cpm. Background is 362 cpm. Then, the lead (%) pigment content is calculated.

Conclusion

Radiochemical methods primarily concern radioactivity study in naturally occurring radioactive substances or other materials in which radionuclides or related compounds are formed by irradiation. Spontaneous electromagnetic radiation or particles emission are monitored from unstable nuclei atoms during radiochemical methods use. Electromagnetic radiation or emitted particles intensity is mainly useful for quantitative analysis, while emission energy is qualitatively analyzed. Alpha particles, neutrons, electrons, positrons, protons, and γ-radiation emission are useful. γ-radiations are actively identical to X rays; although, emitted due to nuclear transformations instead of electron orbital transitions.

Radioisotope assay is done by radiochemical method. It is possible in many cases that non-radioactive sample can be bombarded to temporarily transform a sample part or all into radioactive material with a particle or radiation which can be assayed. Occasionally it is possible for diluting non-radioactive sample along with radioactive isotope to be assayed. To analyze nonradioactive analyte, emission intensity from added radioisotope is measured. This technique is known as dilution analysis of isotope. This chapter provides the comprehensive information about the different methods of radiochemical analysis, their principles and applications; types of radioactive decay; instrumentation and numerous detectors used for the detection of radio-active substance and various application of radio-active substances.

Conflict of Interest

None to Declare.

Consent for Publication

None to Declare.

Acknowledgment

None to Declare.

References

Alimarin, I. P. and G. N. Bilimovitch, The isotope dilution method and its application to analysis of inorganic substances. *The International Journal of Applied Radiation and Isotopes,* 1960. 7(3): p. 169-181.

Buchtela, K., Radiochemical Methods | Overview, in *Encyclopedia of Analytical Science* (Second Edition), P. Worsfold, A. Townshend, and C. Poole, Editors. 2005, Elsevier: Oxford. p. 1-8.

Chapter 1 Introduction, in *Studies in Environmental Science*, H. A. Das, A. Faanhof, and H. A. van der Sloot, Editors. 1983, Elsevier. p. 7-13.

Choppin, G., Jan-Olov Liljenzin, Jan Rydberg, Christian Ekberg. Chapter 18 - Uses of Radioactive Tracers, in *Radiochemistry and Nuclear Chemistry* (Fourth Edition), G. Choppin, G. Jan-Olov Liljenzin, Jan Rydberg, Christian Ekberg. Editors. 2013, Academic Press: Oxford. p. 545-593.

Choppin, G., Jan-Olov Liljenzin, Jan Rydberg, Christian Ekberg., Chapter 5 - Unstable Nuclei and Radioactive Decay, in *Radiochemistry and Nuclear Chemistry* (Fourth Edition), G. Choppin, Jan-Olov Liljenzin, Jan Rydberg, Christian Ekberg, Editors. 2013, Academic Press: Oxford. p. 85-123.

Chu, F. S., Immunoassays | Radioimmunoassay and Enzyme Immunoassay, in *Encyclopedia of Food Sciences and Nutrition* (Second Edition), B. Caballero, Editor. 2003, Academic Press: Oxford. p. 3248-3255.

Duatti, A., Inorganic Radiochemistry, in Reference Module in Chemistry, *Molecular Sciences and Chemical Engineering*. 2019, Elsevier.

Eberle, D. C., 10 - Radioactive tracer methods, in *Techniques for Corrosion Monitoring*, L. Yang, Editor. 2008, Woodhead Publishing. p. 265-276.

Frenzel, W., *Book Review: Principles and Practice of Analytical Chemistry*. 5th Edition. By F. W. Fifield and D. Kealey. Angewandte Chemie-international Edition - ANGEW CHEM INT ED, 2002. 41: p. 1249-1250.

Glascock, M. D., 15.15 - Nuclear Spectroscopy, in *Treatise on Geochemistry* (Second Edition), H. D. Holland and K. K. Turekian, Editors. 2014, Elsevier: Oxford. p. 273-290.

Haba, H., et al., Tracer Technique, in *Handbook of Nuclear Chemistry*, A. Vértes, Zoltán Klencsár, Rezső G. Lovas, Frank Rösch, Editors. 2011, Springer US: Boston, MA. p. 1761-1792.

Hussein, E. M. A., 4 - Transmission, in *Computed Radiation Imaging*, E. M. A. Hussein, Editor. 2011, Elsevier: London. p. 37-52.

Kónya, J. and N. M. Nagy, Chapter 4 - Radioactive Decay, in *Nuclear and Radiochemistry* (Second Edition), J. Kónya and N. M. Nagy, Editors. 2018, Elsevier. p. 49-84.

Kricka, L. J. and J. Y. Park, Assay Principles in Clinical Pathology, in *Pathobiology of Human Disease*, L. M. McManus and R. N. Mitchell, Editors. 2014, Academic Press: San Diego. p. 3207-3221.

L'Annunziata, M. F., 1 - Nuclear Radiation, its Interaction with Matter and Radioisotope Decay, in *Handbook of Radioactivity Analysis* (Second Edition), M. F. L'Annunziata, Editor. 2003, Academic Press: San Diego. p. 1-121.

L'Annunziata, M. F., 3 - Gamma- and X-Radiation — Photons, in *Radioactivity*, M. F. L'Annunziata, Editor. 2007, Elsevier Science B.V.: Amsterdam. p. 187-215.

Loveland, W., Nuclear Chemistry, in *Encyclopedia of Physical Science and Technology* (Third Edition), R. A. Meyers, Editor. 2003, Academic Press: New York. p. 597-615.

Obodovskiy, I., Chapter 16 - Chemical Processes Under the Influence of Radiation, in *Radiation*, I. Obodovskiy, Editor. 2019, Elsevier. p. 247-255.

Ojovan, M. I. and W. E. Lee, 2 - Nuclear Decay, in *An Introduction to Nuclear Waste Immobilisation* (Second Edition), M. I. Ojovan and W. E. Lee, Editors. 2014, Elsevier: Oxford. p. 7-19.

Ojovan, M. I., W. E. Lee, and S. N. Kalmykov, Chapter 2 - Nuclear Decay, in *An Introduction to Nuclear Waste Immobilisation* (Third Edition), M. I. Ojovan, W. E. Lee, and S.N. Kalmykov, Editors. 2019, Elsevier. p. 9-22.

Parry, S. J., ACTIVATION ANALYSIS | Neutron Activation, in *Encyclopedia of Analytical Science* (Second Edition), P. Worsfold, A. Townshend, and C. Poole, Editors. 2005, Elsevier: Oxford. p. 1-10.

PHipps, R. J., Rebecca Jufer Phipps, Jessica Jennings Smith, William D. Darwin, Edward J. Cone, Chapter 2 Current methods for the separation and analysis of cocaine analytes, in *Handbook of Analytical Separations*, M. J. Bogusz, Editor. 2008, Elsevier Science B.V. p. 73-125.

Poston, J. W., Dosimetry, in *Encyclopedia of Physical Science and Technology* (Third Edition), R. A. Meyers, Editor. 2003, Academic Press: New York. p. 603-650.

Steinhauser, G. and K. Buchtela, Chapter 3 - Gas Ionization Detectors, in *Handbook of Radioactivity Analysis* (Third Edition), M. F. L'Annunziata, Editor. 2012, Academic Press: Amsterdam. p. 191-231.

Ter-Pogossian, M. M. and M. E. Phelps, Semiconductor detector systems. *Seminars in Nuclear Medicine,* 1973. 3(4): p. 343-365.

Tölgyessy, J. and M. Harangozó, Radiochemical Methods | Food and Environmental Applications, in *Encyclopedia of Analytical Science* (Second Edition), P. Worsfold, A. Townshend, and C. Poole, Editors. 2005, Elsevier: Oxford. p. 79-86.

Tölgyessy, J. and M. Harangozó, Radiochemical Methods | Natural and Artificial Radioactivity, in *Encyclopedia of Analytical Science* (Second Edition), P. Worsfold, A. Townshend, and C. Poole, Editors. 2005, Elsevier: Oxford. p. 8-15.

Vajda, N., P. Martin, and C.-K. Kim, Chapter 6 - Alpha Spectrometry, in *Handbook of Radioactivity Analysis* (Third Edition), M. F. L'Annunziata, Editor. 2012, Academic Press: Amsterdam. p. 363-422.

Chapter 7

Spectroscopic Methods of Analysis

Malik Saadullah[1*], Muhammad Rashad[1], Shahid Shah[2], Mohammad Alqarni[3], Muhammad Asif[4] and Saeed Ahmad Rao[4]

[1]Department of Pharmaceutical Chemistry, Government College University, Faisalabad, Pakistan
[2]Faculty of Pharmacy, Government College University, Faisalabad, Pakistan
[3]Department of Pharmaceutical Chemistry, College of Pharmacy, Taif University, Saudi Arabia
[4]Faculty of Pharmacy, Islamia University of Bahawalpur, Pakistan

Abstract

Spectroscopy involves measuring the interaction between electromagnetic radiation and the atoms or molecules associated with the sample. Spectroscopy is the scattering of light into components or their respective colors. In simple terms, it is a technique that measures how much light a chemical absorbs and the intensity of the light that passes through it. According to analytical science, each element or compound has a unique characteristic spectrum. Each compound absorbs and scatters discrete light in a specific wavelength range.

Keywords: spectroscopy, analysis, absorption spectroscopy, emission spectroscopy, resonance, FTIR, NMR

---

*Corresponding Author's Email: Maliksaadullah@gcuf.edu.pk.

In: Functional Analytical Techniques in Pharmaceutical Chemistry
Editor: Malik Saadullah
ISBN: 979-8-88697-059-3

Introduction

Spectroscopy is a technique used to express the intensity of radiation measurements which is a wavelength function and usually used to explain methods of experimental spectroscopy. The spectrum measuring equipment is called a spectrometer, spectrophotometer or spectrum analyzer.

Daily color observations may be related to the spectrum. Neon lighting is a direct application of the atomic spectrum. To stimulate emission frequencies of neon and the other noble gases (colors), neon lights employ the phenomena of electrodes crashing with gas. Tin, fibres and paints consist of chemicals chosen in order to generate certain hues and tints according to their spectral properties. Ink, stains and paint are composed of compounds and are selected according to characteristics of their spectrum to produce distinct shades and colors. The mostly used molecular spectrum is the molecular spectrum of nitrogen-dioxide. Nitrogen-dioxide gas, has a special red color absorption characteristic which makes the air polluted by the appearance of reddish brown nitrogen-dioxide. Rayleigh-scattering is a special kind of spectral scattering-phenomenon which is described by the color of the sky. Spectroscopic research, including Max Planck's explanation about the radiation of the darkened body, and Albert Einstein's description of the photoelectric effects and his clarification about the atomic structure and its spectrum, is highly essential to the development of quantum mechanics. In analytical physics and chemistry, spectroscopy is utilized as every atom and molecule has its own unique range. These spectrums are therefore able to be analyzed for the detection, identification and quantification of atomic and molecular information. In astronomy and remote sensing on Earth, spectroscopy is also employed. Most research telescopes have equipment of spectroscopy. The spectrum measured is used to examine the physical and chemical composition of astronomical subjects (temperature and speed).

Theory of Spectroscopy

A resonance and its associated resonant frequency is one of the fundamental concepts of spectroscopy. Mechanical devices such as pendulums, were the first to exhibit resonance. When mechanical systems vibrate or oscillate at their resonance frequency, they will experience large amplitude oscillations. On a graph of amplitude versus excitation frequency, a peak appears centered on the resonance frequency. This graph is a spectrum form with peaks commonly called spectral lines that are comparable in appearance to the most spectral lines. In quantum mechanical systems the equivalent resonance is a pair of two quantum mechanical stationary stations such as an atom linked to a photon by an oscillating precursor of energy. The combined energy between both states is the largest when the energy of the precursor is equal to the energy difference. A photon's energy (E) is proportionate to its frequency (v).

$$E = h\upsilon$$

where h is Planck's constant, resonant frequencies or energy peaks are the response and photon frequency spectra. Particles like electrodes and neutrons can improve resonant interactions, since the connection of de Broglie is significantly interrelated among kinetic energy, wavelength, and frequency. The spectrum of an atom or molecule usually consists of a number of spectral lines with resonances in each line between two distinct quantum states. One of the mysteries of tests which led to the creation and acceptance of quantum mechanics was the description of series of spectral patterns. For the first time, the Rutherford Bohr model of the hydrogen atom has succeeded in drawing the hydrogen spectral series. The spectral lines are clearly separated and identifiable in certain experimental results, but if the energy levels are dense enough they may be nestled and shown in one transition. In theory there are sharp, diffuse and fundamental series for the lines of the given series.

Types of Interaction between Light and Material

Spectroscopy may also be categorized by the type of interaction between energy and the material. The following interactions include:

Absorption Spectroscopy

Absorbance takes place by absorbing the copy energy of the substance from a radioactive source. Absorption measures the percentage of energy delivered through the substance after it has been absorbed which reduces the transferred portion and is often calculated.

Emission Spectroscopy

Represents the radioactive energy emitted by radioactive material. The blackbody spectrum of a substance is a continuous emission spectrum measured at temperature. Infrared rays, such as a copy interferometer emitted in the atmosphere can measure this feature. Other energies like fires, sparks, electric arcs or electromagnetic copies in the event of fluorescent, fluorescence can also help to reduce emissions.

Elastic Scattering and Reflection Spectroscopy

Measures how the test material reflects or disperses the radiation. Scattering of high energy such as X-rays and electrons is used for cryptography to study the structure of proteins and the atomic arrangement of solid crystals.

Impedance Spectroscopy

A characteristic of a medium that interferes or slow down the energy transfer. For optical applications, it is distinguished by the index of refraction.

Inelastic Scattering Phenomena

It can change the wavelength of scattered radiation according to the energy exchange between the radiographically inspected materials. These include the dispersion of Raman and Compton.

Coherent Or Resonance Spectroscopy

A technique where radiation energy combines two quantum states into a coherent, radiation-field-supported interaction of a material. Consistency can be recorded through particle collisions and other interactions such as transmission of energy, which generally needs a high level of radiation. Spectroscopy with atomic magnetic resonance (NMR) is a commonly utilized approach. In infrared and visible areas, ultrafast laser spectroscopy may also be conducted.

Nuclear Spectroscopy

A technique that uses specific nuclear properties to investigate matter, the most commonly condensed matter, the local structure of liquid or frozen liquid and biomolecular elements.

Type of Material

Spectroscopic studies explain the interaction of radiant energy with certain types of matter.

Atoms

Atomic Spectroscopy is a pioneering development application. Visible and ultraviolet light is used in atomic absorption spectroscopy and atomic emission spectroscopy. These absorption and emission of nuclear spectral lines often occur as electronic transitions of external shell electrons rise and fall from one electron orbital to another. Other atoms also have a clear X-ray spectrum that is recognized to bring internal bark electrons into an excited state. Atoms of different elements have a specific spectrum, so atomic spectroscopy can be used to identify and quantify the elemental composition of a sample. After inventing the spectrometer, Robert Bunsen and Gustav Kirchhoff used this technique to observe the emission spectrum and discover new elements. The atomic Hupsson is found in the spectrum of the Sun and is called the Fraunhofer line. The thorough explaination of the range of hydrogen described the lamb shift which was detected in the spectrum of hydrogen and a success for both electro-mechanics in its early stages. Modern applications of atomic spectroscopy are flame spectroscopy, induced plasma spectroscopy, glow discharge spectroscopy, atomic emission spectroscopy causes inductively coupled plasma, spark or arc spectroscopy for the study of visible and ultraviolet transitions. X-ray spectroscopy is a technique used to study the X-ray spectrum of X-ray fluorescent molecules.

Molecules

Atoms bond to molecules creating discrete types of energy states and thus the transition between these energy levels is linked to a unique spectrum. Molecular spectra may be produced via molecular rotation, molecular vibrations and electronic states according to electron spin (electron paramagnetic resonance). In the microwave and in millimeter wave areas, collective motion which is known as oscillations in atomic nuclei mainly contributed to create spectras. Rotational spectroscopy is a synonym of microwave spectroscopy. Vibrations are caused by the relative motion of atomic nuclei and are generally studied in both Raman and infrared spectroscopy. Studies including fluorescence spectroscopy, visible and ultraviolet spectroscopy, require electron excitation. Molecular spectroscopic research gives a path to the invention of lasers which lead to further advancement of lasers.

Crystals and Extended Materials

Atoms or molecules into a crystal or other expanded form results in additional energy states/ levels. Since these states are multiple, their density increases. This high density often creates a weaker, less specific spectrum. For example, thermal motion of atoms and molecules in matter induces blackbody radiation. Acoustic and mechanical consequences also occur in collective movements. However, particular spectrum transitions can occur in pure crystals and the sequences identified also influence the investigation of molecular spectrum. The uniform grid structure disperses rays, electrons or neutrons to facilitate crystallographic investigations.

Nuclei

Nuclei also have several energy states that result in a wide-range of gamma radiation spectrums. Certain nuclear spin states can separate magnetic field energy levels which enables spectroscopy of nuclear magnet resonance feasible.

UV-Visible Spectroscopy

Ultraviolet-visible spectroscopy in the UV-visible region and the entire area of the neighbouring visible spectrum introduces absorption spectroscopy and reflecting spectroscopy. This means the use of visible or nearby light. The visible range immediately influences the sensing color of the related chemical via the absorption or reflection of the light. Atoms and molecules undergo electronic transitions in electromagnetic spectrum. Absorption spectroscopy is critical for fluorescence spectroscopy, as fluorescence transitions are from excited to ground state and absorption leads to a transition to excited ground again. A flow chart of UV-visible spectroscopy is illustrated diagrammatically in Figure 7.10.

Introduction

The interaction of electromagnetic waves with matter at specified energy level is based on spectrophotometry. The spectrophotometry generally speaking, to measure or interpret electromagnetic waves absorbed into the sample as molecules moving from one energy condition to another (i.e., from ground to excited (or finished) state then to ground state.

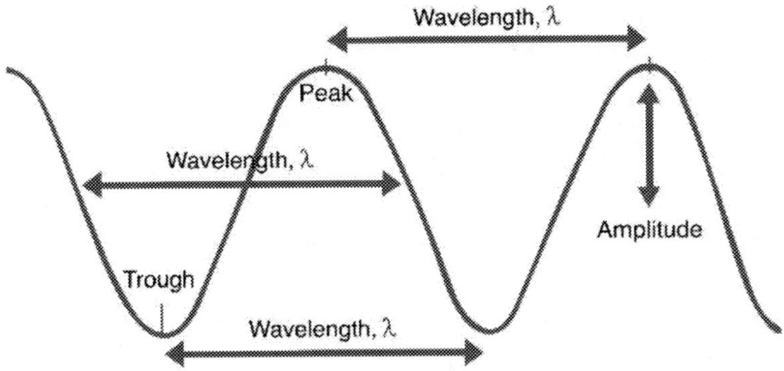

Figure 7.1. Wave phenomenon.

Electromagnetic waves are generally expressed in terms of wavelength (λ), i.e., the distance between two successive peaks; wave number (v), i.e., the number of waves per centimeter and frequency, i.e., the number of waves per second (Figure 7.1).

Waves Phenomenon

The mathematical relationship among these three quantities is expressed by the following:

$$c = \lambda v \qquad \text{(i)}$$

Law of quantum mechanics may be applied to photons as

$$E = hv \qquad \text{(ii)}$$

where E is the energy of the radiation; n is the frequency; and h is Planck's constant.

Combining equation (i) and (ii)

$E = hc/\lambda$

In the visible region, wavelengths are in units of nanometers (nm), i.e., 10-9 m, but can also be expressed in other units, such as millimicrons (μm) and angstroms (Å).

1 nanometre = 1 nm = 1 mμ = 10 Å.

The visible spectrum is generally considered to be 380-770 nm, and the UV region is generally defined at 200-380 nm. Light absorption in both the UV and visible light regions of the electromagnetic spectrum occurs when light is facilitated by vibrational or rotational transitions, similar to the spectrum required to generate electronic transitions from molecules.

Figure 7.2 below describes a simple UV/visible absorption spectrum for buta-1,3-diene. Absorbance (on the vertical axis) is a measure of the amount of light absorbed. The higher the value, the more the wavelength is being absorbed. Figure 7.3 illustrates the electromagnetic spectrums.

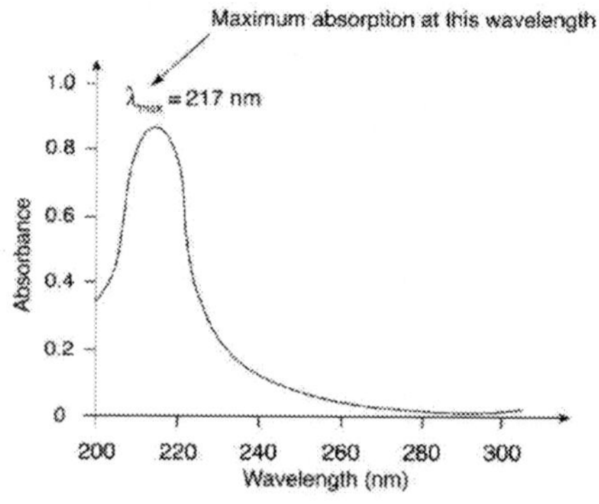

Figure 7.2. Absorption maxima diagram.

Radiation type	Wavelength, λ (Å)	Frequency $\nu = c/\lambda$ (Hz)	Applications
Radio	10^{14}	3×10^{4}	Spin orientation
Nuclear magnetic resonance	10^{12}	3×10^{6}	
Television	10^{10}	3×10^{8}	
Radar	10^{8}	3×10^{10}	
Microwave	10^{7}	3×10^{11}	Rotational
Far infrared	10^{6}	3×10^{12}	Vibrational
Near infrared	10^{4}	3×10^{14}	
Visible	8×10^{3} - 4×10^{3}	3.7×10^{14} - 7.5×10^{14}	Electronic
Ultraviolet	3×10^{3}	3×10^{15}	
X-rays	1	3×10^{18}	
Gamma rays	10^{-2}	3×10^{20}	Nuclear transitions
Cosmic rays	10^{-4}	3×10^{22}	

Figure 7.3. Table for the electromagnetic spectrums.

Theory

Ultraviolet visible (UV-visible) spectroscopy is an important key to obtain absorption spectra of compounds as solutions or solids themselves. Spectroscopy is the absorption of electromagnetic radiation or light energy and is used to excite electrons in a compound or material from an elementary level of energy to a single, excited state. The UV-visible electromagnetic spectral energy area comprises 1.5-6.2 eV, which corresponds to an 800-200 nm wavelength. The Law of Runbelt Bale is a fundamental principle underpinning spectroscopy of absorbance. For a single wavelength, the length of the cuvette track of the sample holder (typically 1 cm), and the concentration of the solution is C, the molar extinction coefficient of the compound selected is ε of solution (M).

$$A = \varepsilon bc$$

All of these instruments have a radioactive light source (usually a deuterium lamp or a tungsten lamp) and a sample holder attached to the detector but some instruments use only one wavelength filter at a time. A monochromator between the light source and the analyst can be used for single beam devices to examine a wavelength at a time (Figure 7.4). The double beam instrument (Figure 7.5) comprises a beam splitter and a light-free light source, and a set of mirrors which are used to concentrate and evaluate the beam on the reference sample to provide more precise reading. In contrast, there is no monochromator or filter between the sample and the source on the synchronous instrument (Figure 7.6) but there is a diode array detector which allows the instrument to measure absorbance at all wavelengths. Synchronized instruments are generally faster as well as more efficient than other instruments although types of these spectrometers work adequately.

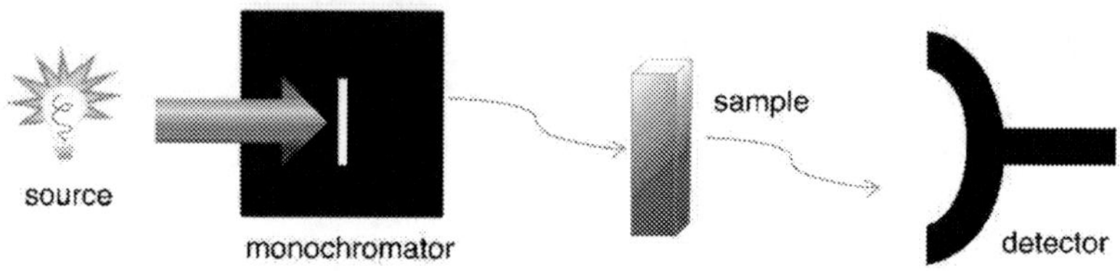

Figure 7.4. Illustration of a single beam UV-visible instrument.

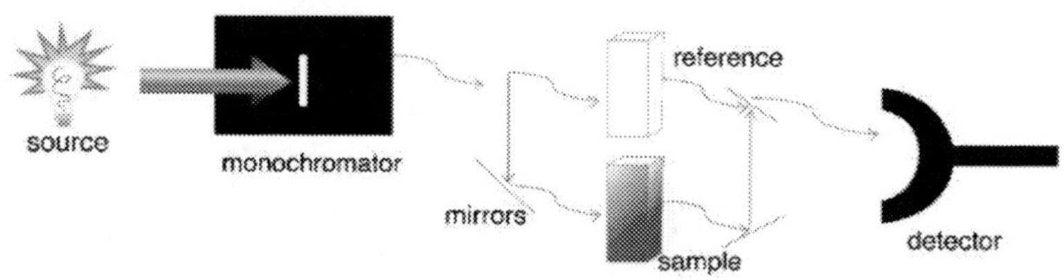

Figure 7.5. Illustration of a double beam UV-visible instrument.

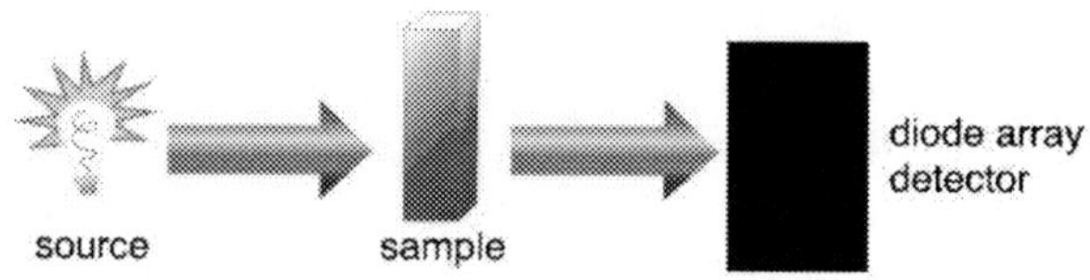

Figure 7.6. Illustration of a simultaneous UV-visible instrument.

Absorption of Different Electromagnetic Radiations by Organic Molecules

In ultraviolet, infrared and nuclear resonance areas, spectra of absorption undergo many processes but a certain energy quantity is the fundamental process. The energy necessary to get from low energy (E1) to high energy (E2) is identical to the electromagnetic radiation energy that generates the transition (Figure 7.7).

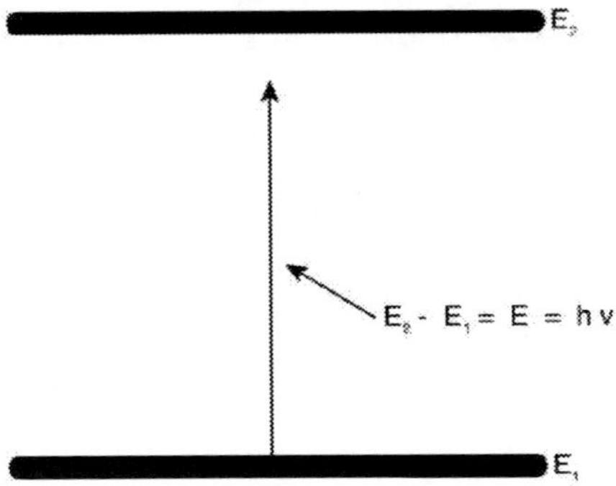

Figure 7.7. Absorption of Different Electromagnetic Radiations by Organic Molecules.

Electronic Transitions

Ultraviolet or visible light absorption leads to external electrons being excited. Electronic transitions of three kinds exist, for instance:

1. Transitions involving $p(\pi)$, $s(\sigma)$, and n electrons.
2. Transitions involving charge-transfer electrons.
3. Transitions involving d and f electrons

Electrons travel from the ground level to the excitation level, when they are absorbed by atoms or molecules. The atoms can spin and vibrate in a molecule accordingly. These vibrations and rotations also have special levels of energy as indicated in the Figure 7.8, that may be stated at the top of every electron energy level.

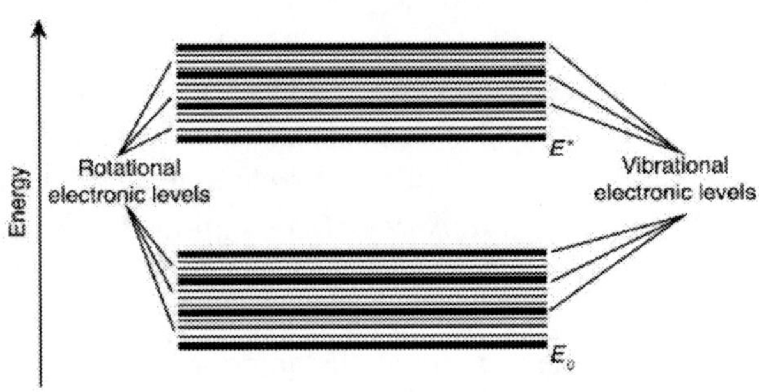

Figure 7.8. Electronic levels diagram.

Absorbing Species Containing π, σ, and n Electrons

Ultraviolet and visible organic molecular light absorption is restricted to particular functional groups of low energy excitation (chromophores). These chromophores have a complicated molecular spectrum. This is because the electronic transition overlaps oscillatory and rotational transitions leading to overlap across line combinations. There is hence a permanent absorption band.

Possible electronic transitions of π, σ, and n electrons are the following (as expressed in Figure 7.9):

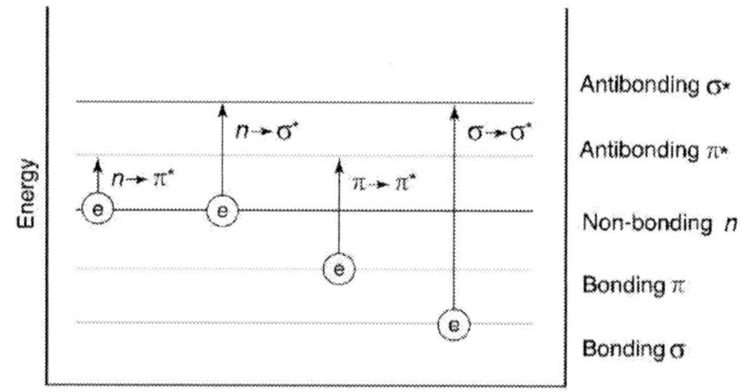

Figure 7.9. Types of transitions diagram.

- *σ σ* Transitions:* The electrons in the bonding orbitals move to their respective semi bonding orbitals hence greater energy is required. For example, methane (which may give σ σ* with only C-H bonds) shows a peak absorbance at 125 nm. Absorption peaks due to the σ σ* transition are not observed in the typical UV, visible light spectrum (200-700 nm).
- *n σ* Transitions:* Saturated compounds with atoms with isolated pairs (unbounded electrons) exhibit n σ * transitions. These transitions typically require less energy than σ σ* transitions. It can generate light with wavelengths in the range of 150 to 250 nm. The number of organic functional groups with n σ* peak in the UV region is small.
- *n to π* and π to π* Transitions:* Most absorption-spectroscopy of organic compounds relies on the transition of n or p electrons to the p* excited state. This is because the absorption peaks of these transitions lies in the experimentally appropriate spectral region (200-700 nm). These transitions require unsaturated groups in the molecule

to donate p- electrons. Molar absorbance from n π* transition is relatively low and belongs to 10-100 L mol^{-1} cm^{-1}.

The transition to π* from π generally provides a molar absorbance of 1000- 10,000 L mol^{-1} cm^{-1}. Absorbent Paper in which solvent that dissolves also affects the spectrum of the species. At n, the peak moves to a shorter wavelength (blue shift) with higher solvent polarity at the π* transition. This is because the plume of the lone pair of electrons increases and the energy of the n orbital decreases. Sometimes (not always), the opposite (i.e., redshift) is observed in the π to π* transition. This degrades the two energy states, the excited state and the non-excited state due to the polarizing force of the attraction between the solvent and the absorber. This effect is because of the excited state is higher, resulting in a slight decrease in the energy difference between the excited and unexcited states and a small red method deviation.

This effect also causes a transition from n to π*, but is masked by the blue shift due to the plum blossom of the lone pair of electrons.

Beer–Lambert's Law

When "*C*" is in moles, the constant is called molar absorptivity (or) molar extinction coefficient *(ε)*:

$$A = \epsilon CT$$

ε can also be written as:

$$\varepsilon = E_{1\,CM}^{1\,\%} \times \frac{molecular\ weight}{10}$$

where $E_{1\,CM}^{1\,\%}$ *is the absorbance of 1 % w/v solution using a path length of 1 cm.*

Instrumentation

The different components are the following:

1. Radiation source

2. Monochromators
3. Sample cells
4. Detector
5. Recorder (or) display.

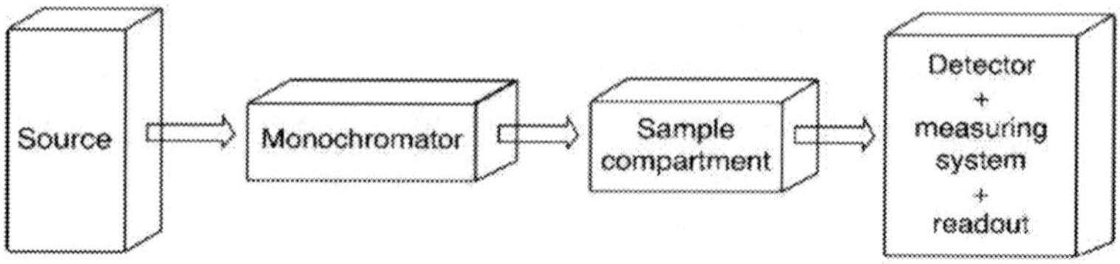

Figure 7.10. Flow chart of instrument of UV spectroscopy.

Radiation Source

Tungsten–Halogen Lamp

This lamp produces light with wavelengths up to near-ultraviolet (300-320 nm) at the red end (750-800 nm) of the visible spectrum. The lamp is coated with a quartz rind which enables UV radiation to be used.

Hydrogen (or) Deuterium Lamp

This is usually used to measure UV light far away (up to 200 nm). There are two electrodes submerged in deuterium-packed silica. Figure 7.11 shows the figure for deuterium hydrogen (or) light and a tungsten halogen lamp.

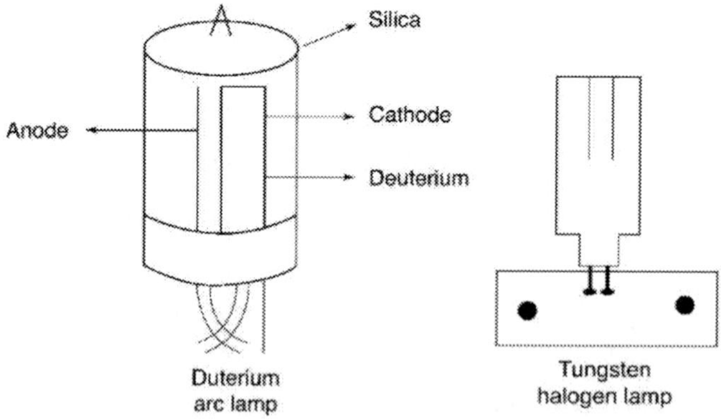

Figure 7.11. Lamps diagrams.

Monochromator

The role of the monochromator is to allow white light to be used over a wide wavelength range and to divide it into individual wavelength copies. The main components are (a) inlet slits, (b) parallel devices for producing parallel light, (c) wavelength selection or in a dispersion system, (d) focal lens or mirror, (e) exit slit, as shown in the Figure 7.12.

Two basic methods of wavelength selection may be used: filters and a dispersing system (e.g., a prism or diffraction grating).

Filters

Colored glass or gelatin is very limited in its use because the simplest form of filter is limited to the visible region and has a wide spectral bandwidth. Typical bandwidths are often recommended over 30-40 nm. Filters are basically composed of materials with different indices of refraction deposited on a substrate (generally glass, but possibly silica) with a bandwidth of 10 nm or less.

Prisms

Prism of the appropriate material and shape gives the wavelengths of separated components a continuous spectrum. The incidence slit (to define the incident beam) and collimator is used to better define light between the source and the prism (to generate a parallel beam at the prism). The prism usually spins to shift the spectrum across the output split. The following figure 7.12 shows a conventional prism monochromator. UV systems are preferred mirrors for efficiency and economic reasons rather than reflecting components i.e., lenses.

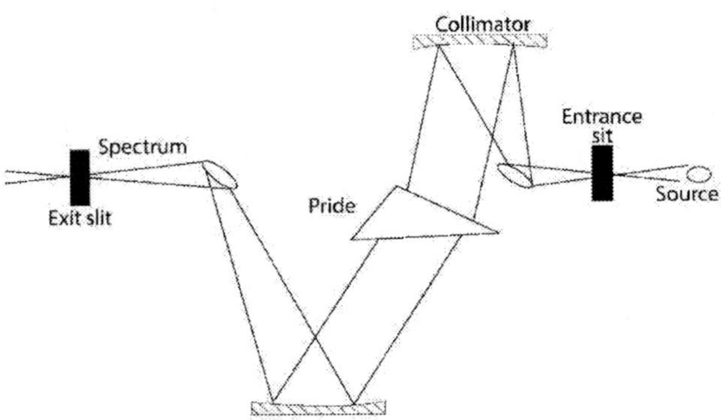

Figure 7.12. Simple condenser prism monochromator.

Prism monochromators with UV / visible bandwidths of 1 nm and above are effortlessly achieved, resulting in achieving significant performance compared to filter-based designs. However, using prisms has the following drawbacks: non-linear dispersion, temperature-related properties of commonly used prism materials, complex prism drive mechanisms needed to provide convenient wavelength control and measurements.

Diffraction Gratings

The grating is an alternate way that monochromatic light is produced (Figure 7.13). A diffraction grid is essentially a sequence of parallel grooves (lines) that is reproduced by a master meticulously manufactured using a machine or by a hologram-generated master. It will be configured. The home preferentially selects or selects the wavelength to be reflected when the reflected light interacts with the light reflected by the adjacent home to generate interference and the angle of the grid of the incident beam is changed. It may be considered another mirror. Compared to prisms, gratings offer a higher resolution linear dispersion and thus a simpler mechanical design for constant bandwidth and wavelength selection as shown in Figure 7.14.

The relationship that determines the wavelength of the reflected light is expressed by the following:

$$n\lambda = 2d \sin \theta$$

where n is in the order, d is the distance of the reflected light and θ is the angle of the radiation incident. The rotation of the gate in the beam alters the direction and chooses the reflection of the wavelength.

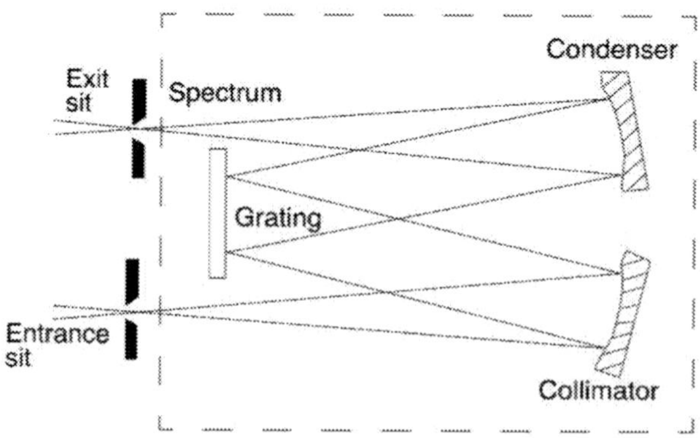

Figure 7.13. Grating's monochromator.

Two additional characteristics of gratings may be noted:

1. If wavelength λ is reflected for a given angle q, then λ/2, λ/3 and so on are also reflected at the same angle.
2. The energy and the resolution of a grating is directly related to the number of lines. For maximum efficiency, the line separation should be one wavelength, and for UV/visible gratings, the line density should be 1200 per mm.

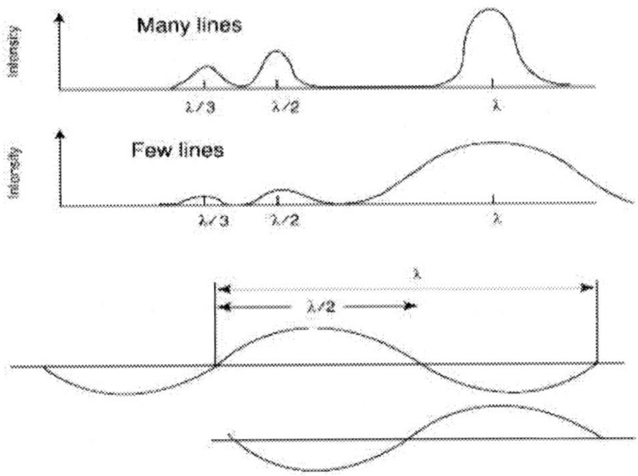

Figure 7.14. Efficiency of the gratings.

Gratings have the following advantages over prisms:
Better resolution and energy transfer.

1. Linear dispersion and therefore constant bandwidth.
2. Less complicated wavelength drive mechanism is required.
3. Stray light is limited to imperfections at the grating surface.

Optical Geometry

These measurements are made using cuvettes with the appropriate solvent poured in the light path. When measuring absorption, it is essential to re-measure the reference at every specific wavelength of the measurement.

Single-Beam Optics

The advancement of microprocessors enabled the improved outcomes to be achieved with individual beams compared to twin beams. Optical and mechanical simplicity is improved.

Split (Reference) Beam Optics

A xenon flash lamp was used in the spectrophotometer, which required a split beam configuration. Because in xenon pulsed lamps, high-brightness flashes do not always have to be the same size. In a black and white meter, about 70% of the energy passes through the sample.

Double-Beam Optics

Conventionally, the most common method is a double-beam geometry in sample processing. Double-beam operation is provided by a time-division system in which the optical path alternately passes through the sample and reference cells. After startup, the UV/visible spectrophotometer automatically generates the following absorption spectrashown in the below Figure 7.15.

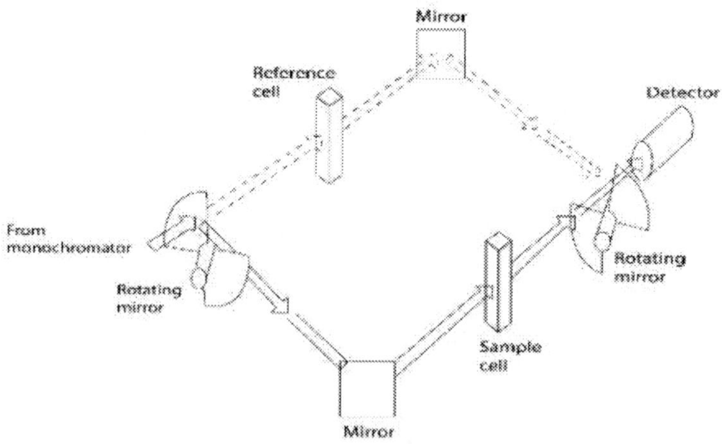

Figure 7.15. Arrangement of the instrumental components for double beam optics.

Diode Array Optics

The fourth optical arrangement is the monochromatic diode array after light passes through the sample, such that the sample compartment lid is not necessary. Most importantly, the grid is fixed and not working like previous systems.

Sample Cells

The sample for analysis can be in the solid, liquid or gaseous state, and the substance containing the sample must be transparent at the measurement wavelength (Figure 7.16).

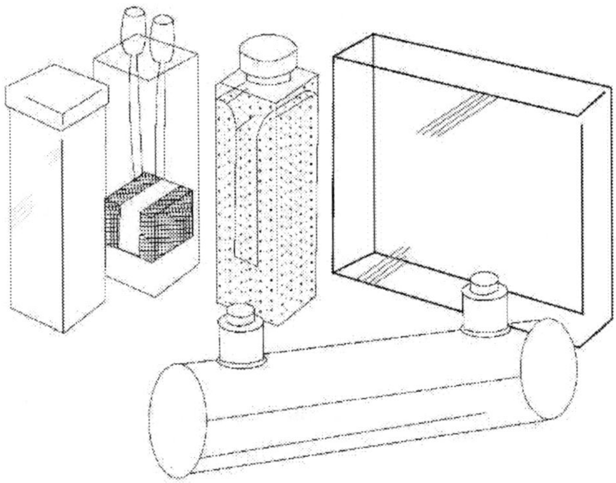

Figure 7.16. Illustration of sample cells.

Cells composed of optically flat molten glass are used for measurements in the UV / visible region above 320 nm, transparent to less than 180 nm below 320 nm, consist of expensive molten silica. The cells of molten silica with path length 10 nm are used for normal procedures and 1-50 nm for special applications.

Detectors

The photomultiplier tube is widely used in UV/visible light spectrometers (Figure 7.17). Multiplication of light by secondary emission is intended to get higher sensitivity for feeble light intensities. Multiple anodes with continuously increasing potentials are used in a single bulb.

The electrons in the photocathode move to anode 1 and move to anode 2 to generate more electrons that continue to provide 106-108 times the final current than the primary current in anode.

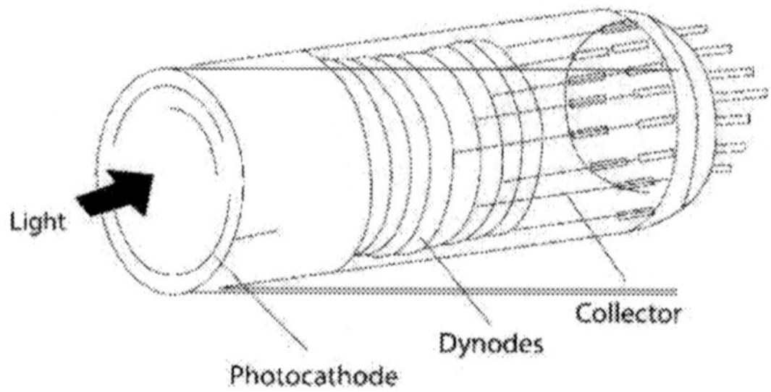

Figure 7.17. An illustration of photomultiplier tube detector.

Silicon Diode

Silicon diode detectors perform better than photomultiplier tubes and have a wider wavelength range but may be drift and drop. Electronic benefits include mechanically a solid-state device, decreased supply of power and control system needs.

Recorders

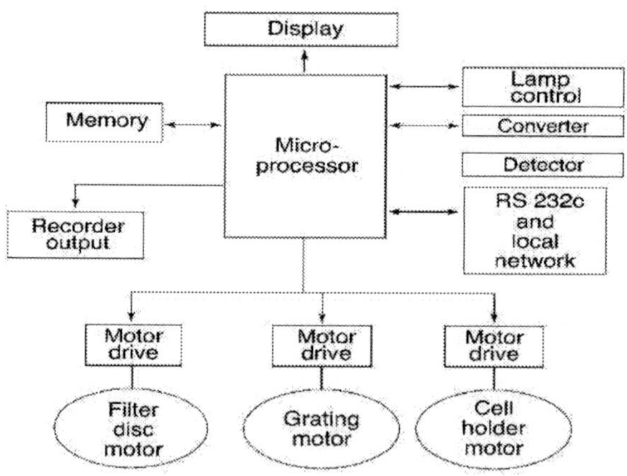

Figure 7.18. Block diagram of a microprocessor-controlled spectrophotometer.

Spectrophotometer ends with the result of a signal (voltage) associated with sample absorption at a given wavelength. The meter provides the output signal absolute value and in certain cases, the point of the board of the backoff circuit. Digital measurements (LEDs or LCDs) are used due to lack of clarity and ambiguity. For example, they are connected to microprocessors that read directly in concentration units. A block schematic diagram of the electronic processing of the post-detection unit for the integrated outcomes and control of a sophisticated single-beam spectrophotometer is depicted in Figure 7.18.

Limitations of UV-Visible Spectroscopy

Sample

UV-visible spectroscopy works well with liquid samples, but when the liquid sample contains suspended solid particles, the sample particles scatter more than they absorb light and the results are extremely distorted. Most UV visible devices can use a diffractometer (Figure 7.19), to analyze solids or suspensions but they are not commonly used. UV-visible instruments typically analyze liquid samples and solutions more efficiently.

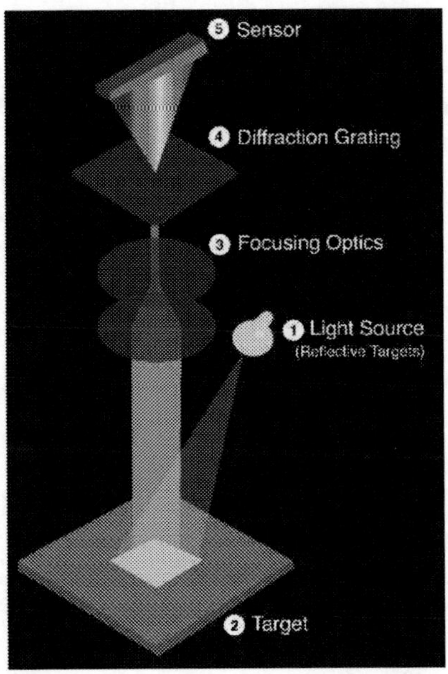

Figure 7.19. Schematic representation of the apparatus for collecting UV-visible spectra from solid materials.

Calibration of Instrument

When starting the analysis of the solvent used (water, hexane, etc.), a desktop reference is required, and if concentration analysis is required, the calibration solution must be accurately prepared. When the solution for the sample is not properly developed, the actual sample concentration cannot be correctly determined.

Choice of Solvent or Container

All solvents have a wavelength value for the UV-visible absorbance block. The solvent-cutoff value is shown below at which the solvent absorbs all the light itself. Therefore, when choosing a solvent, it is necessary to consider the absorbance cutoff and the position where the compound being investigated is likely to be absorbed. If it is close to the cutoff value, select another solvent as sown in Table 7.1.

Table 7.1. Absorbance cutoffs of various common solvents

Solvent	UV Absorbance Cutoff (nm)
Acetone	329
Benzene	278
Dimethylformamide	267
Ethanol	205
Toluene	285
Water	180

Also, a visible UV absorption cut-off is available to the substance making up the cuvette (sample container). The quartz sample holders are more adapted to absorb UV since the sample supporting glass material absorbs high energy from approximately 300 nm, as the quartz absorbance cutoff is around 160 nm (Table 7.2).

Table 7.2. Three different types of cuvettes commonly used, with different usable wavelengths

Material	Wavelength Range (nm)
Glass	380-780
Plastic	380-780
Fused Quartz	<380

Concentration of Solution

To reliably recover data, analyte absorption peak must be at least 3 times the strength of the instrument's background noise. This can be countered by the use of a greater concentration in solution. If the sample is very tiny and the dilution cannot produce an adequate signal, sample holders are present in standard cuvettes that store sample sizes less than 2.5 ml. Certain cuvettes are just 100 μl in size and are used for analysis of tiny samples without significant dilutions, which results in a poorer signal-to-noise ratio.

IR – Spectroscopy

The infrared portion of the electromagnetic spectrum is used by IR spectroscopy (short for infrared spectroscopy). That is light with a larger wavelength than visible light but with a lower frequency (Figure 7.20 shows different regions of light). Infrared spectroscopy generally involves the analysis of molecular-infrared interactions. IR spectroscopy can be analyzed in three ways: reflecting, emitting, and absorbing. The primary purpose of infra-red spectroscopy is to identify organic and inorganic molecular functional groups. Infrared spectroscopy uses the concept that a molecule absorbs a particular frequency of a particular light into the corresponding structure of the molecule. Molecules can absorb energy from incident light and, in return, offer faster rotation or more pronounced vibrations.

Theory

IR spectrum is a graph composed of the Y-axis infrared absorption frequency or the X-axis wavelength.

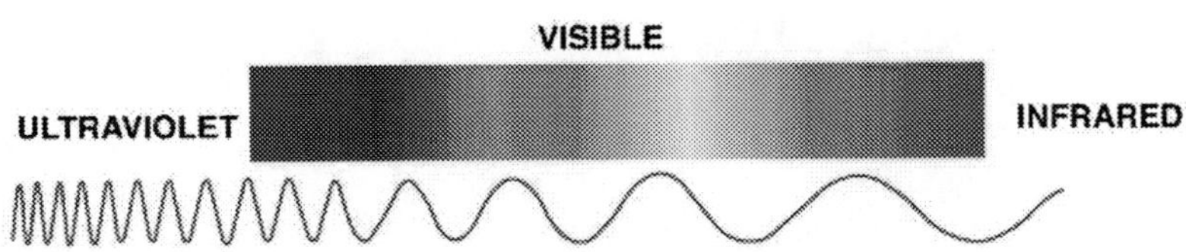

Figure 7.20. A diagram describing the different regions of light.

IR spectroscopy detects the sample molecules absorbing infrared radiation frequency. The molecule absorbs these unique frequencies of light and corresponds to the bound vibration frequency of the molecule. The energy required to vibrate the bonds connected to the molecule here and with greater amplitude is generated in the infrared region. The connections interact only when they are polar in nature with electromagnetic and infrared duplicates. The electric field component of the electromagnetic wave distinguishes the vibration energy of molecule in the presence of some positive and negative charges. Changes in vibrational energy cause other changes corresponding to the dipole moment of the molecule. Absorption strength performs a function as polarity of the bond shows. Nonpolar bonds with O = O and N ≡ N symmetry do not absorb radiation because they cannot interact with an electric field.

Instrumentation

Infrared spectroscopy instrumentation is described below. First, the IR light from the light source splits into two parts and passes through the reference and sample as shown in Figure 7.21.

The two beams pass through the next detector reflected by the splitter. Finally, after processing the data that has passed through the detector, its output shows the required measurements.

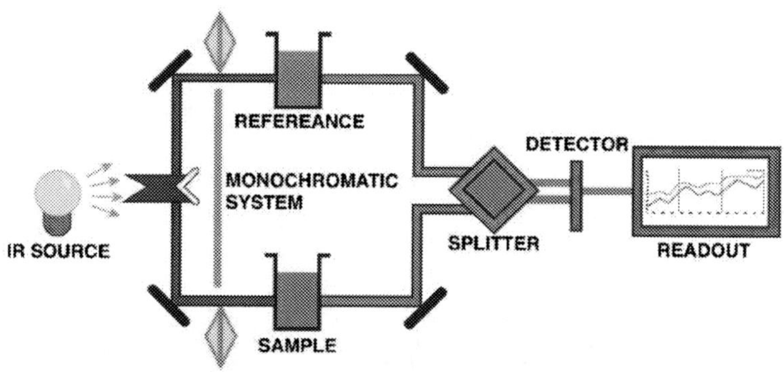

Figure 7.21. An illustration of IR spectroscopy working.

IR Spectrophotometer Consists of Following Components

Radiation Source

Radiation source is an integral part of the IR-spectrophotometer. It continuously produces infrared rays of sufficient intensity. The copy strength must be constant in the λ range and should include a wide range of λ and must be same for a long period of time. The operating temperature of the source is usually 1100-1500 K, the maximum intensity should be 4000-400 cm^{-1}. To reduce noise, the source must be enclosed with an insulator. Modern radiation sources have important examples such as furnace ignitions and diesel engine heaters. Important IR radiation sources include:

The infrared source must be an inert solid which is heated electrically to the required temperature (1,500-2,200 K). After reaching the required temperature, the heated material emits infrared radiation.

The Nernst Glower

The Nernst glower has the shape of a hollow cylinder composed of rare earth oxides such as zirconium oxide, cerium oxide and thorium oxide. A platinum metal lead at the end of the cylinder in the Nernst glower allows the passage of electric current. The resistors are high, negative and must be pre-heated to become conductible. They have high electric resistance.

The Glow Bar Source

The Glow Bar is comprised of a rod with a diameter of 5 mm and 50 mm that needs to be heated to around 1500 K electrically. In order to prevent arcing, electric connections should be cooled down using water. With the exception of a shorter wavelength (less than 5 mm), the spectral output is virtually identical to those of the channel list.

The Carbon Dioxide Laser

An infrared source i.e., gas phase of tuning carbon dioxide laser monitoring certain contaminants of the air and the species absorption in aqueous solution.

Sample Holder

The sample cell contains the sample and permits the sample for spectral IR measurement to pass monochromatic light produced from the light source. It is probable that a sample is examined as a solid, liquid or gas.

It is often used as NaCl or KBr alkaline halide. These halides are usually water-soluble. Aqueous solvents cannot be utilized in the study to dissolve the sample cells when they consist of alkaline halides. Only chemical solvents like carbon disulfide (CS_2) and carbon tetrachloride are analyzed (CCl_4).

Monochromator

The function of the monochromatic machine is to use prisms (NaCl prisms and/or metal halides) and/or grids to divide multicolor radiation into individual wavelengths. The resolution of the IR spectrum depends on the quality of the mirror and the slit width. Rock salt prisms made of metal with a polished surface are commonly used in the range of 650-4000 cm^{-1}.

Detectors

The sensor is an optical signal received from a sample cell which is converted to an electrical signal and inserted into a graph.

Types of Detectors

The three types of detectors are available depending on the type of the chemical composition that monitors the spectrum and the material used to build the detector. You can categorize it into.

Thermal Detectors

The thermal sensor can serve as a function at room temperature extensively in a variety of wavelengths. The primary limitations in comparison with other types of detectors are low reaction time and inadequate sensitivity.

Thermocouple

The detector consists of a junction of several metals which have been fused together. For instance, at both ends of antimony, two bismuths are binding. The electrical difference between the connections relies on the difference in the temperature between the connections. A thermocouple is formed when several thermocouples are connected in series.

Bolometer

The bolometer changes resistance when heated. It includes different metal strips like platinum or nickel, or semiconducting metals.

Pyroelectric Detectors

The pyroelectric sensor comprises of a pyroelectric substance, a thermal and electrical insulator. The most often utilized substance in pyroelectric infrared detectors is triglycine sulphate. Unlike other IR thermal detectors, the pyroelectric effect is not reliant on the detector's temperature but rather on the rate of temperature changes in the detector. The pyroelectric detector then operates at a relatively fast response time, making this detector suitable for Fourier spectrometers that require a fast response.

Photoconducting Detectors

The photoelectric detector is a most sensitive detector. The interaction between photons and semiconductor metals is based on them. It is a thin semiconductor material film detector. The substance is mercury cadmium telluride, lead sulphide or Indium antimonid, which is implanted in a non-conductive surface of glass and sealed in a vacuum bag to prevent ambient conditions. This substance is not available on all materials.

In the near-infrared region of the spectrum, lead sulphide detectors are employed. The detector employs infrared and remote infrared rooms with mercury cadmium telluride. To minimize disruptions, liquid nitrogen should be utilized and is cooled at room temperature.

Read-Out Device

It receives the electrical signals from the detector and translates it into easy to interpret form.

FT-IR - Fourier Transform Infrared

Fourier transform infra-red spectroscopy is a unique infrared spectroscopy technique called FTIR. Infrared radiation has been devised to overcome the sluggish scanning constraints achieved by the disperser using FTIR. The measured signal is sent in grammes. A Fourier transformation on this signal results in a spectrum comparable to existing (dispersed) infrared spectroscopy, although in seconds rather than in minutes, the technology is significantly quicker.

Atomic Spectroscopy

Atomic spectroscopy is an electromagnetic or mass spectrum measurement of the elemental composition. The study of an element's electromagnetic spectrum is known as optical atomic spectroscopy. Electrons are present at different levels of energy in atoms, as illustrated in figure 7.22. These levels have a definite energy to travel, and the electrons must absorb or release energy that is equivalent to their energy difference. Optology demonstrates that when an electron travels to a lower energy state, the energy received to excite an electron into an energy level or the energy released takes the form of a photon.

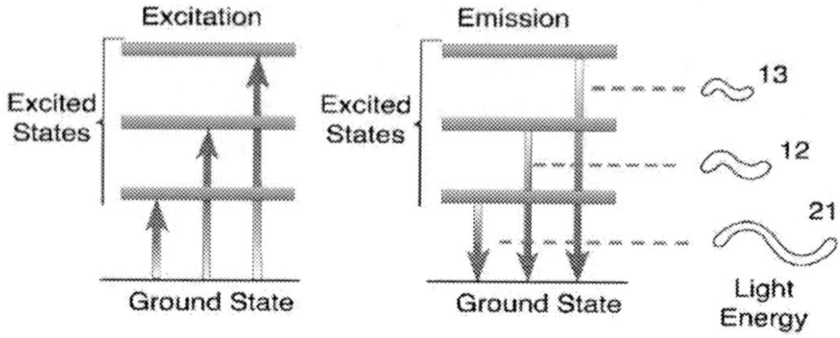

Figure 7.22. Energy transition states.

In the electronic transitions, the wavelength of the emitted radiation is exactly proportional. As the wavelength of the produced light is a unique feature for each element as every element has a distinct electrical design. Other electron transitions can occur because the orbital composition of large atoms can be complicated. Each conversion emits the light having a specific wavelength.

Techniques for Analysis

Atomic spectroscopy has three methods used for analytical purposes. (1) Atomic absorption, (2) atomic emission, and (3) atomic fluorescence. In all three areas of atomic spectroscopy, ground excitation and decomposition processes occur. The energy received during the excitation process is measured and used for test purposes or energy released during the excitation phase.

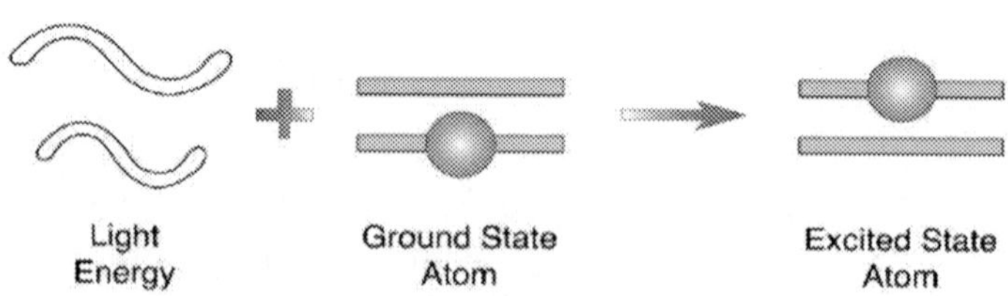

Figure 7.23. The process of atomic absorption.

In atomic absorption, the light of a specific-wavelength is absorbed by an atom and excited from a ground energy-state to a higher energy-state, then amount of absorbed energy is measured (Figure 7.23).

Atomic emissions produce high-energetic, generally thermal, atoms in an excited state ready to release light. Atomic emissions electric arc, fire, or lately plasma can be the energy source. Atomic emission is a third branch of nuclear spectroscopy fluorescence. There are two features in this technique: atomic absorption and atomic emission.

Atomic Absorption Spectroscopy

Atomic absorption spectroscopy (AAS), a spectroscopy method, absorbs electromagnetic light at a certain free gas atom wavelength in order to provide the appropriate quantifiable signal. The absorption signal is linked to the free atom concentration of the optical path.

Instrumentation

It consists of following instruments.

Atomizer

The analyzed sample must be sprinkled first. This is a key stage in AAS for reading sensitivity measurement. A high number of uniform free atoms are produced by the most effective nebulizer. Different kinds of atomizers are available, however, typically only 2 are utilized, one flammable atomizer and one electrothermal atomizer.

Flame Atomizer

Flame atomizers are extensively used for several reasons including easy, cost-effective and extended duration. The aerosol from the atomizer volatilizes the sample and sprays it into a flame with sufficient energy to spray. The sample is then dried, vaporized, sprayed, and ionized. For this type of atomizer, other subcategories are determined depending on the chemical composition of the flame. The flame is mostly composed by the analysis of the sample. Flames have a number of criteria including energy, long duration, no turbulence and safe operation (Figure 7.24).

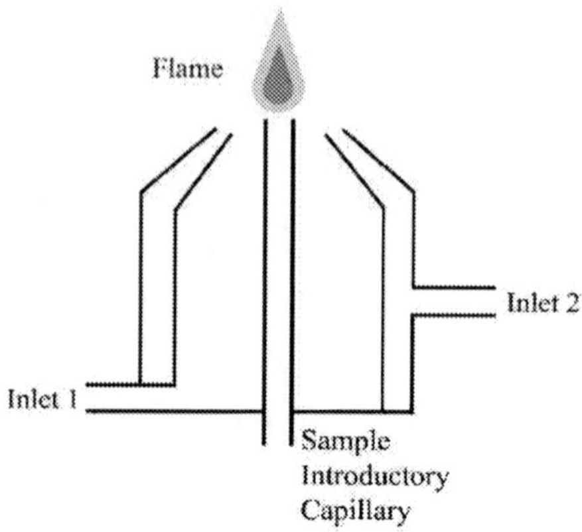

Figure 7.24. A schematic diagram of a flame atomizer showing the oxidizer inlet (1) and fuel inlet (2).

Electrothermal Atomizer

Electrothermal atomizers are not popular since their sensing level has been increased before the flame atomizers. They contain graphite tubes that raise the temperature by way of a bevel. Electrospray evaporates the solvent and impurities after drying and atomize the sample and then elevate the temperature to clean the graphite-tube. The capacity to keep the temperature consistent while spraying is essential for the form spray. Electrothermal nebulizers should be able to atomize fast, maintain an adequate solution volume and produce minimum radiation (Figure 7.25).

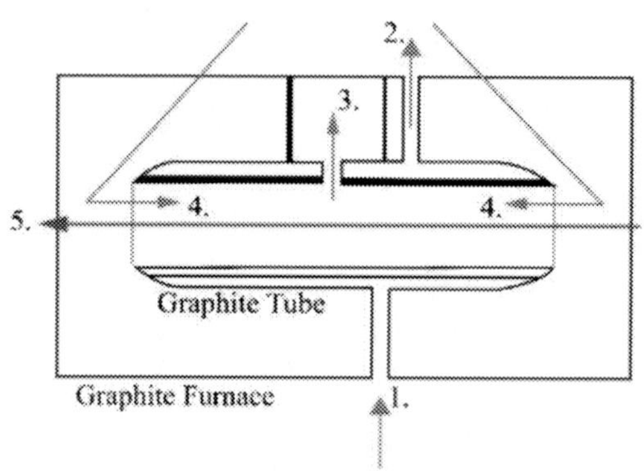

Figure 7.25. Schematic diagram of an electrothermal atomizer showing (1) an external gas flow inlet, (2) an external gas flow outlet, (3) an internal gas flow outlet, (4) an internal gas flow inlet, (5) and the light beam.

Radiation Source

The source of radiation emits the atomized sample matching light. The probe absorbs certain wavelength radiation, while the remaining light flows through the sensor. Two broad kinds of source of radiation exist: sources and rings of continuum. The line source pushes the analyte through the sample and emits its own range. Examples of widely used sources include hollow cathode lights and electrode discharge lamps. On the other hand, continuous sources generate radiation which stretches over a greater area. These sources are often used to adjust the backdrop. Deuterium and halogen lights are also often employed.

Spectrometer

Spectroscopy is used to divide light into multiple wavelengths before it is passed to the detector.

AAS spectrometers can be single-beam or double-beam spectrometers. Single-beam spectrometers require radiation to pass through an atomized sample, whereas double-beam spectrophotometers require two beams. One source goes through the sample and another goes through the reference. The schematic diagram of the double beam spectrophotometer is shown below in Figure 7.26.

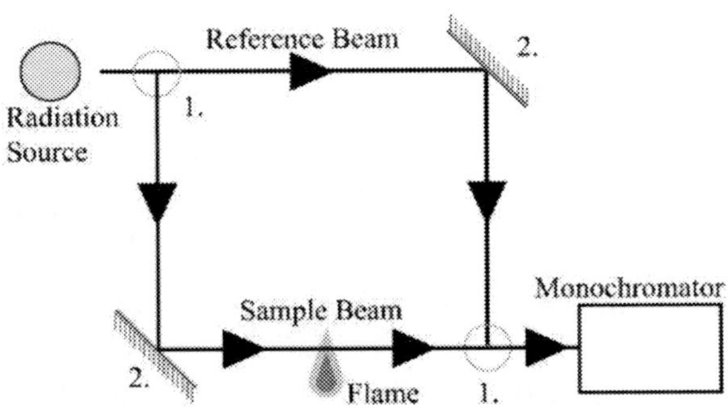

Figure 7.26. A schematic diagram of a double-beam spectrometer showing (1) the 50/50 beam splitters and (2) mirrors.

Calibration Curve

Use a calibration curve to measure the concentration of the analyte from the solution. Absorbance concentration plots are created using the standard. The three widely used methods for creating calibration curves are bracketing techniques, standard calibration techniques and additional techniques to the analyte.

Atomic Emission Spectroscopy

Atoms excited to a high energy state move to a low energy state by emitting radiation (luminescence or emission). Excitation caused by high temperature energy, emission is called atom or light emission (atomic emission spectrum), and excitation caused by light is called atomic fluorescence (atomic fluorescence spectrum).

Atomic Emission Spectroscopy (AES) uses quantitative measurements of atomic light emission to determine the concentration of analytes in a sample. The analyte atoms in the solution move to an excited state, where they are desolvated, vaporized, and atomized by flame, electric discharge, or plasma. These high temperature atomization sources provide large amounts of energy to excite atoms to their higher energy levels. The atoms return to a lower level of light emission, as described in Figure 7.27.

Principle

The design of atomic emission spectroscopy is similar to the design of atomic absorption spectroscopy. However, most flame atomic absorption, not hollow cathode lamps, employ atomic absorption spectrophotometers to detect the energy difference released by the sample heated in a flaring position above the flame. The largest number of atomic emission spectrometer are instruments intended for the major atomic emission functions comparable to the usage as atomization sources of plasmas, arcs, sparks and lasers.

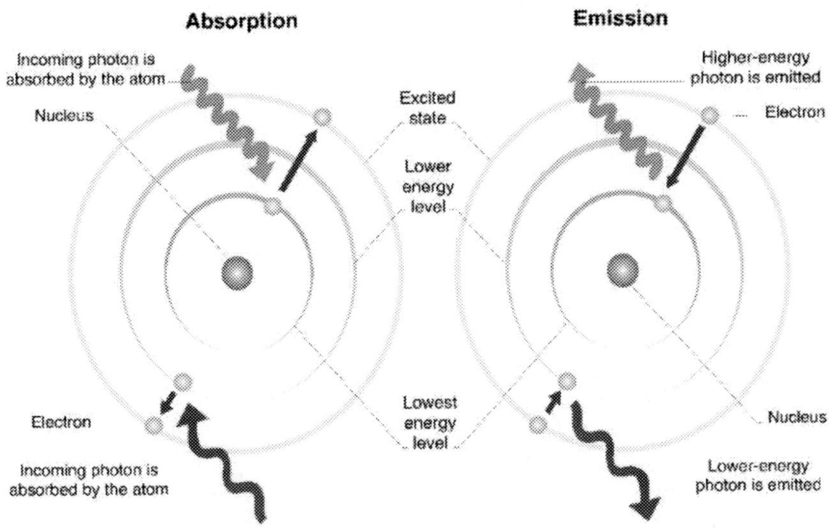

Figure 7.27. An illustration of atomic emission and atomic absorption phenomenon.

Instrumentation

Atomic emission spectroscope equipment consists of a spray such as a flame or plasma source and a detector connected to the source here, a monochromator and a reader.

Atomization and Excitation Source

The conversion of an analyte from a solid, liquid, or solution into gaseous atoms in glass is needed for atomic emission spectroscopy. The most common methods of atomization and excitation are flame and plasma.

Flame Sources

This technique uses a flame to dispense and pump a spray of a sprayed sample in an enclosed chamber. A flame burner can contain multiple slots. Previous instrumentation uses atomization and a flame source.

Plasma Sources

Plasma is a partly ionized, heated gas with many good ions and electrons. Plasma ionizes an argon gas flow that produces argon ions and generates electrons. Since plasma performs functions at higher temperatures than fires, they are more atomizing and give more population in stimulated countries. The inductively coupled plasma contains a sample aerosol consisting of three concentric silica quartz tubes, in which argon gas passes through a central tube and the argon gas is ignited by a flame from a Tesla coil. As a result, the atoms of the analyte are excited to high energy levels. Figure 7.28 below shows a schematographic representation of a plasma source inductively coupled (ICP).

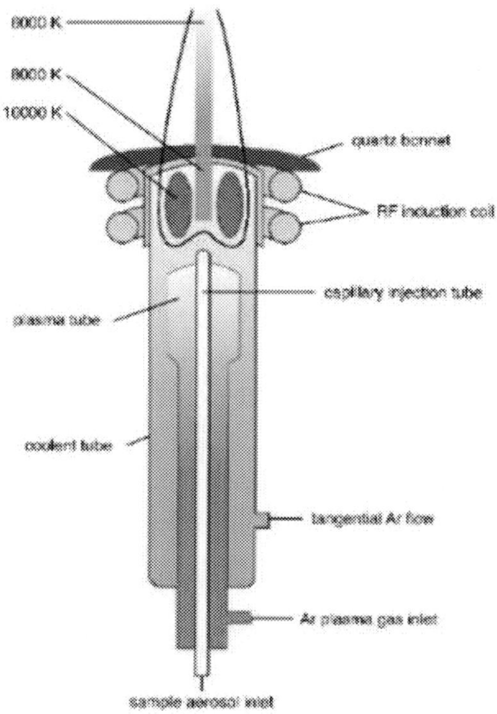

Figure 7.28. An inductively coupled plasma torch.

Monochromator

Monochromator divides the ribbon into different wavelengths and selects the wavelength necessary. A prism or grating is used as a monochromator.

Detector

The monochromator allows it to pass through a detector that filters the required wavelengths and converts the optical signal into an electrical signal. The wavelengths of the sampling may be detected using the detector. A photomultiplier tube, a sensitive vacuum excursion is the most widely utilized detector at AES.

Readout Device

A microprocessor-controlled electronic device is used for the reader. The device converts the signal detected by the detector into the same readable format as a visual graph or print format.

Mass Spectroscopy

Mass spectroscopy is a technique of analysis for measuring the mass to charge ratio (m/z) for a sample of molecules. This approach is used to measure the molecular precision of a component of the sample. Generally, mass spectrometers may be utilized through molecular weight measurements to identify new chemicals, quantify known compounds and evaluate the sample structure and chemistry of the molecules.

Principle

Mass spectrometry relies on the generation of positive ions. The technique is that the sample is first converted to a vaporized form, and then high-energy electrons (>10 eV) are bombarded to create immediately charged ions called molecular ions or parent ions that have electrons removed from the metal as shown in Figure 7.29.

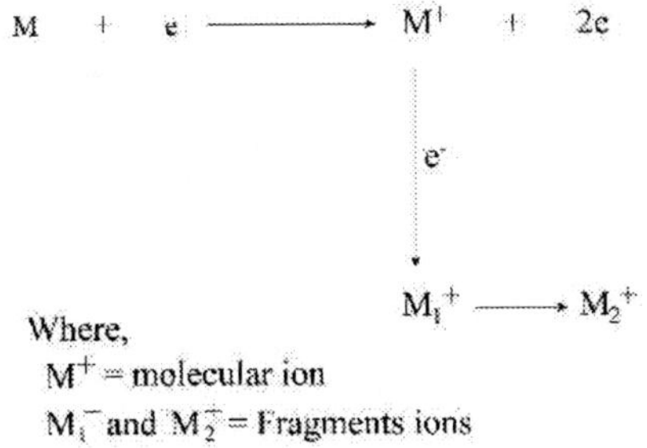

Figure 7.29. Ionization of molecule by electron bombardment.

The actively charged molecule M+ is usually unstable and decomposes into fragments or auxiliary ions denoted M+1 at energies (10 to 70 eV) depending on the bond strength. The formed ions are subjected to the forces of electric and magnetic fields, resulting in a mass spectrum recorded by a detector separated from the analyzer.

Components of Mass Spectrometer

Mass spectrometer consists of following three components (Figure 7.30):

1. Ionization Source
2. Mass Analyzer
3. Ion Detection System

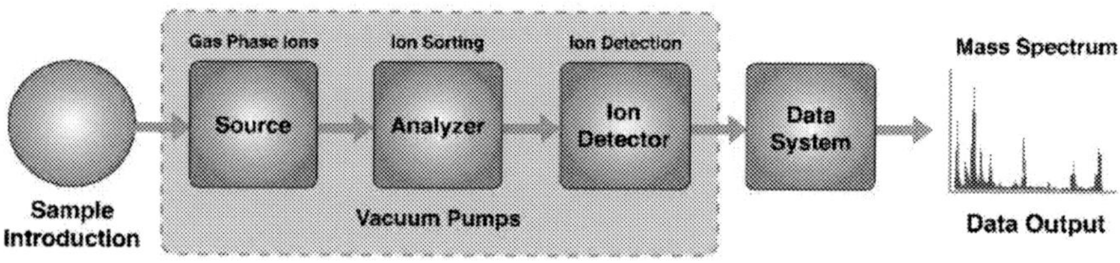

Figure 7.30. A schematic diagram of mass spectrometer.

Ionization Source

The sample is transferred to the next ionization chamber that was first vaporized. The electrically heated metal coil pulls towards the electron trap, which is a positively charged plate and emit electrons. A particle (atom or molecule) in a sample collides with a series of electrons which ares powerful enough to knock one or more electrons in the sample to create a cation. Already formed cations are almost +1 charged because it is already very difficult to remove additional electrons with cations. This cation is moved to the rest of the machine by an ion repulsion plate, which is a slightly positively charged metal plate. The ion sources for the mass spectrometer are shown in Figure 7.31.

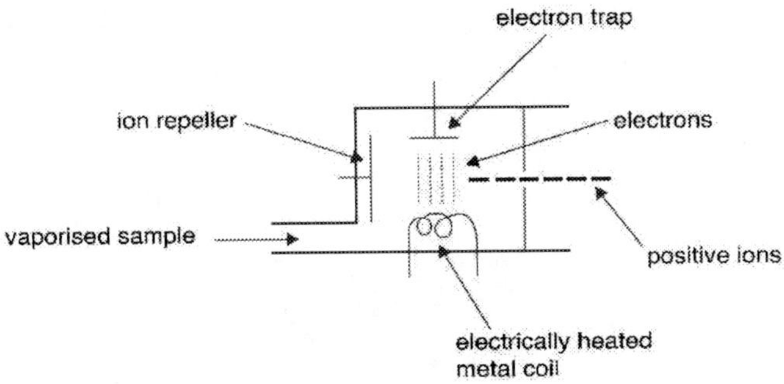

Figure 7.31. Ionization source of mass spectrometer.

Acceleration

The positive ions are accelerated and fall out of the high positive ionization box and pass through three slits. The middle slit carries medium voltage and the last slit carries 0 volts. After that, the ions will be accelerated to finely focused beam (Figure 7.32).

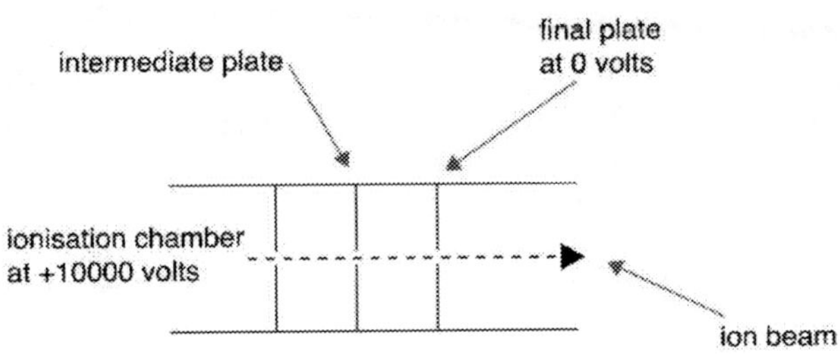

Figure 7.32. Acceleration plates.

Deflection

Ions are deflected by the magnetic field by other amounts based on their mass. The deflection amount relies on the following terms:

1. Mass of ions. Those that are more deflected are lighter ions.
2. Ion charges are more deflicted than ions with a positive load, i.e., ions with two (or more) positive loads.

The mass-charge ratio is based on these two parameters. In certain situations this ratio is marked with the sign m/z (or m/e). For instance, if an ion weighs 28 and has a load of 1+, its load ratio is 28. Ions of weight with 56 and a load of 2+, with a weight/load ratio is 28.

The most inclusive stream A of the ions in the Figure 7.33 is because they have the least weight/charge ratio of ions. Ion stream C contains ions with the maximum mass / charge ratio, so it is the minimum biased value.

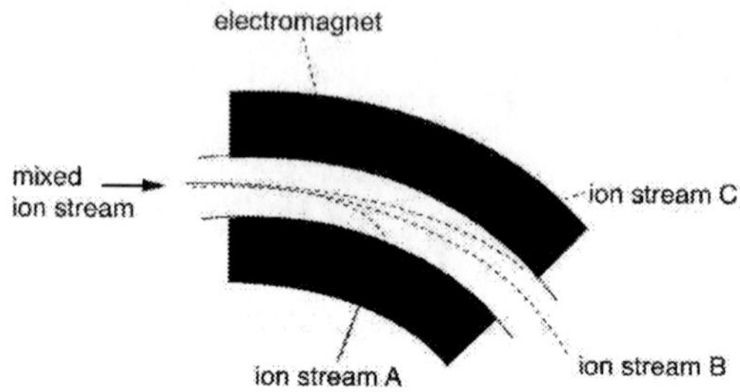

Figure 7.33. Deflection plates.

Detection

Only ion stream B is scented and moves to the right with the ion detector. The remaining ions are neutralized by colliding with the wall and retaining the electrons. Finally, it was removed from mass spectrometry with a vacuum pump. When ions collide with a metal box, the metal is neutralized by the ions with electrons (Figure 7.34). It leaves a space between the metal electrons, and the wire electrons are competing to fill it.

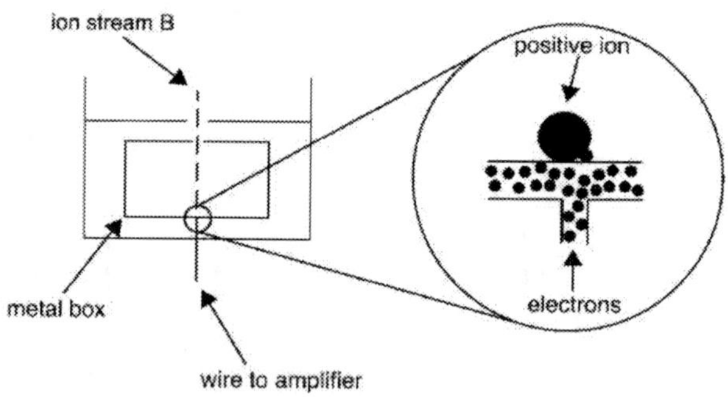

Figure 7.34. A diagram showing detection of charge.

The electric stream that passes across the wire is detected, amplified and recorded as an electrical current. The greater the current, the more ions flow. The mass spectrometry figure is the following illustration in Figure 7.35.

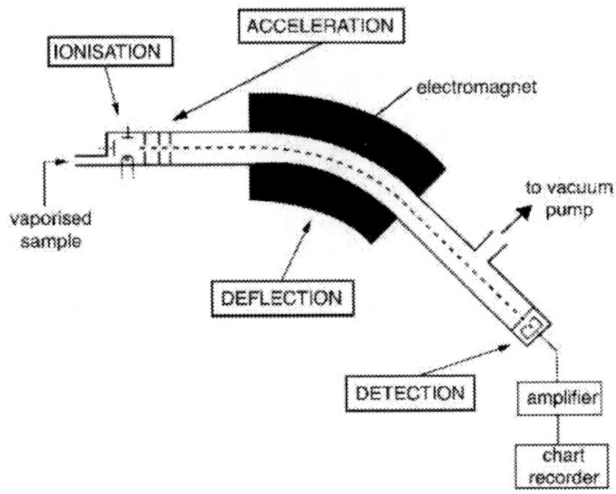

Figure 7.35. An illustration of mass spectroscopy.

NMR Spctroscopy

(NMR) Nuclear magnetic resonance spectroscopy is a type of spectroscopy which detects the local magnetic field surrounding the nucleus. It is also known as magnetic resonance spectroscopy or MRS. This is a spectroscopic method that relies on the nuclei to absorb electromagnetic waves in the radio frequency domain with a wavelength of 4 to 900 MHz. NMR has been the primary method for identifying the organic chemical structure in the past 50 years. For NMR spectroscopy, a full evaluation and interpretation of the whole spectrum is usually anticipated, compared to other spectroscopy. An illustration diagram of NMR spectroscopy instrumentation is shown in Figure 7.40.

Principle

The principle of NMR is that almost all atomic nuclei are charged with atomic nuclei spin. In applying an external magnetic field, energy transfers from the condition of the ground to the higher energy level. Transfering energy takes place at a wavelength relative to a radio frequency, and the energy is released at the same frequency if the spin returns to its basic state. This energy is measured, processed and produced in a number of ways by producing NMR spectra from the atomic nuclei. The main points are:

1. All nuclei are charged and have many nuclear spins.
2. With an external magnetic field, energy passes to higher energy levels.
3. Energy transfer takes place at a radio frequency matching wavelength.
4. Furthermore, energy is released at the same frequency when the spin returns to its basic state.
5. Therefore, processing of the NMR spectrum of the nucleus concerned can be obtained by measuring a signal similar to the transmission.

Spin and Magnetic Properties

Atomic nuclei contain particles called neutrons and protons using a unique property called spin. Like electrons, nuclear spins are identified by quantum numbers so that I is used for spins, and m is used for spins in magnetic fields. There are even protons and neutrons at nuclear spins at zero, and strange atoms have a nonspin. This spin leads to the nucleus acting as a tiny magnet bar. Each spin magnet is randomly oriented in the absence of an external magnet field (Figure 7.36).

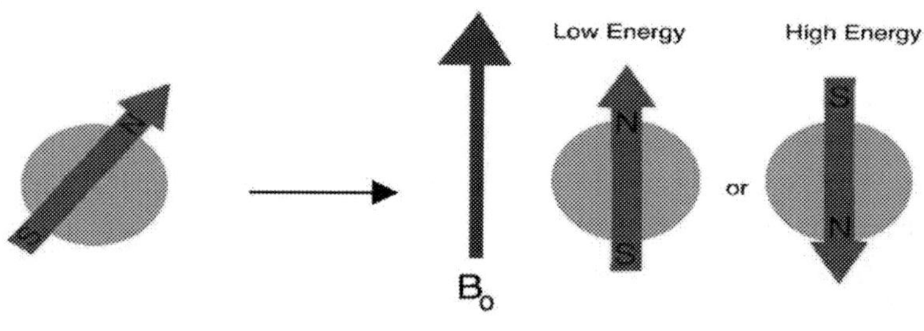

Figure 7.36. A diagram showing that when an external magnetic field is applied the spin is oriented.

Spin Flip

Spin flips between energy levels occur for the NMR instrument to operate. The difference in energy between the two states influences the electromagnetic energy level in the nucleus. In most NMR spectrometers, B0 is represented by Tesla (T) and γ is represented by 107. Generally, the required electromagnetic copy is 100 MHz and GHz or higher. The energy of photons is explained as follows:

$$E = hv$$

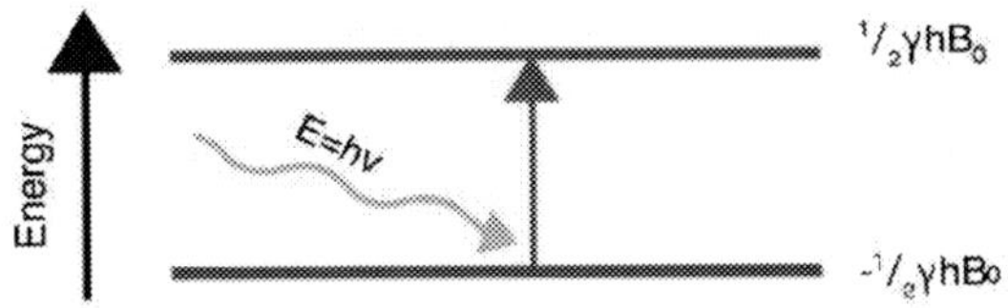

Figure 7.37. Absorption of radiation.

Radiation absorption in order to produce a shift from nuclear energy levels is termed as spin reverse (Figure 7.37).

Nuclear Shielding

The effectiveness of the NMR hinges on the notion to provide for the structural assignment of nuclear shielding. A wide range of electrons around the nucleus are present on every atom. Charged particles can produce the magnetic field sensed by the nucleus such as electrons that move in the loop. Therefore, inducing a small change in the next level of energy in the electromagnetic environment around the core is a slight change in the magnetic field of the nucleus. This is known as a shield and is expressed in Figure 7.38.

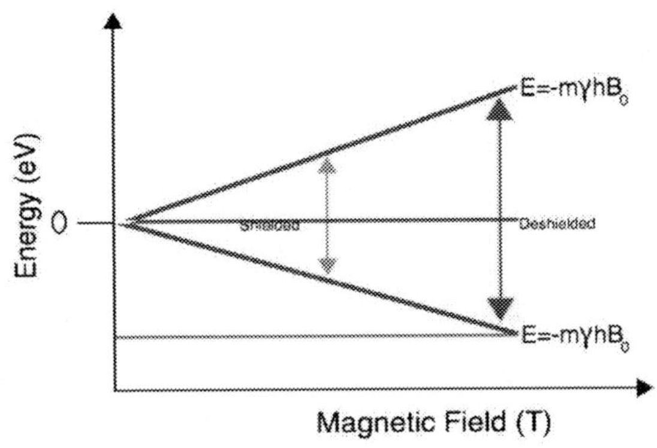

Figure 7.38. Nuclear shielding process.

Relaxation

Relaxation refers to a phenomenon in which the nucleus of an atom is excited to a higher energy level and then returns to a thermodynamically stable state, as illustrated in Figure 7.39. The energy used by an atom is generated and released as it changes from a lower level of power to a greater level of energy.

This can be a difficult process for various rest periods. Relaxation in spinning grids (T1) and spinning spins are the most popular kinds of relaxation (T2).

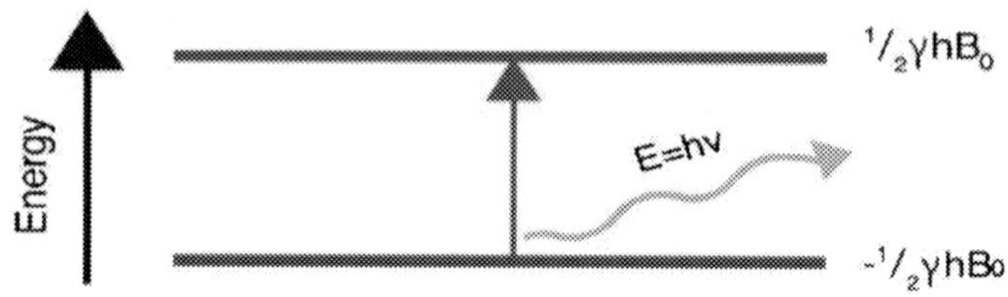

Figure 7.39. The process of relaxation to emit radiations.

Spin-Lattice Relaxation (T1)

It is a time to recover from the X-Y plane along the Z axis for a 37 percent mass magnetization. The easier you relax, the lower the time of relaxation (T1). Motions between molecules are limited in solids; hence the value of relaxation time (T1) is high. The measurements of spin-lattice are usually measured by pulse techniques.

Spin-Spin Relaxation (T2)

T2 is the time needed for spinning to lose coherence. T2 may be less or equal to T1.

Instrumentation

This instrument consists of nine major parts. They are discussed below:

1. Sample holder – Sample is held by a glass tube which is 8.5 cm long and 0.3 cm in diameter.
2. Magnetic coils – Whenever current flows through it, the magnetic coil generates magnetic current.
3. Permanent magnet – It helps to provide a homogenous magnetic field at 60 – 100 MHZ
4. Sweep generator – It modifies the magnetic field strength which is already applied.
5. Radiofrequency transmitter – It produces a short and powerful pulse of radio waves.
6. Radiofrequency – It helps to detect receiver radio frequencies.
7. RF detector – It helps to determine the unabsorbed radio frequencies.

8. Recorder – It records the signals of NMR which are received by the RF detector.
9. Readout system – A computer that shows and records the data.

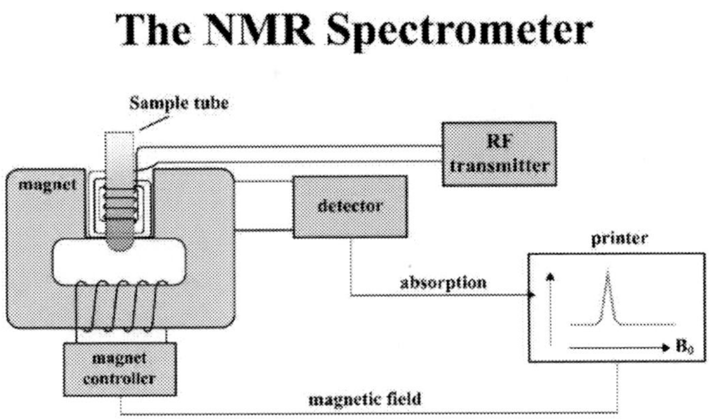

Figure 7.40. An illustration diagram of NMR spectroscopy instrumentation.

Conclusion

Spectroscopy measures electromagnetic waves of different frequencies, from radio waves to X-rays that are absorbed, scattered, or emitted by an analyte under appropriate conditions. There are many different types of radiation that can be monitored, and the methods of measuring radiation make up the largest category of spectrometric methods because they can vary widely from method to method. Spectroscopy can be mainly divided into three categories: UV and visible absorption spectroscopy, fluorescence spectroscopy and infrared (IR) spectroscopy. The last is nuclear magnetic resonance (NMR) spectroscopy. This chapter briefly explains the principles of spectroscopy and related equipment.

Conflict of Interest

None to Declare.

Consent for Publication

None to Declare.

Acknowledgments

None to Declare.

References

Beckett A, Stenlake J (1997) *Practical Pharmaceutical Chemistry, part II, vol 1*. CBS Publications and Distributors, New Delhi, pp 275–300.

Beckett A, Stenlake J (1997) *Practical Pharmaceutical Chemistry, part II, vol 1*. CBS Publications and Distributors, New Delhi, pp 275–300.

Fernández B, Lobo L, Pereiro R (2019) Atomic absorption spectrometry. Fundamentals, instrumentation and capabilities. In: Worsfold P, Poole C, Townshend A, Miró M (eds) *Encyclopedia of Analytical Science, 3rd edn*. Academic, Oxford, pp 137–143.

Fifield FW (2000) *Principles and Practice of Analytical Chemistry*. Blackwell, Oxford.

Fifield FW (2000) *Principles and Practice of Analytical Chemistry*. Blackwell Science, Hoboken.

Förster HUV (2004) Vis spectroscopy. In: Karge HG, Weitkamp J (eds) *Characterization I Molecular Sieves – Science and Technology, vol 4*. Springer, Berlin, pp 337–426.

Gauglitz G (2000) Ultraviolet and visible spectroscopy. In: *Ullmann's Encyclopedia of Industrial Chemistry*. Wiley, Weinheim.

Gauglitz G, Dakin JP (2017) Spectroscopic analysis. In: John PD, Robert GWB (eds) *Handbook of Optoelectronics, vol 2*. CRC Press, Boca Raton, pp 569–600.

Gauglitz G, Dakin JP (2017) Spectroscopic analysis. In: John PD, Robert GWB (eds) *Handbook of Optoelectronics, vol 2*. CRC Press, Boca Raton, pp 569–600.

Gauglitz G, Moore DS, Vo-Dinh T (2014) Handbook of spectroscopy. Wiley, Hoboken.

Gauglitz G, Moore DS, Vo-Dinh T (2014) *Handbook of Spectroscopy*. Wiley, Hoboken.

Hollas JM (2002) Basic atomic and molecular spectroscopy. *Royal Society of Chemistry*, London.

Kemp W (2017) Organic spectroscopy. *Macmillan International Higher Education*, London.

Labmate Online. *Atomic Absorption Spectroscopy*. Accessed 12 Sept 2019.

LibreTextsTM. *UV-VIS Spectroscopy*. Accessed 12 Aug 2019.

LibreTexts™. *Introduction to Spectroscopy*. Accessed 2 Sept 2019.

Pavia DL, Lampman GM, Kriz GS, Vyvyan JA (2014) *Introduction to Spectroscopy*. Cengage Learning, Boston.

Pavia DL, Lampman GM, Kriz GS, Vyvyan JA (2014) *Introduction to Spectroscopy*. Cengage Learning, Boston.

Schrenk W (2012) *Analytical Atomic Spectroscopy*. Springer, Berlin.
Sneddon J (2002) *Advances in Atomic Spectroscopy*. Elsevier, Amsterdam.
Sudha PC (2012) *Pharmaceutical Analysis*. Pearson Education India, Chennai.
Sudha PC (2012) *Pharmaceutical Analysis*. Pearson Education India, Chennai.
Waters Corporation. *Introduction to Spectroscopy*. Accessed 10 Aug 2019.
Waters Corporation. *UV-VIS Spectroscopy*. Accessed 15 Aug 2019.
Watson DG (2015) *Pharmaceutical Analysis E-book: a textbook for pharmacy students and pharmaceutical chemists*. Elsevier Health Sciences, Amsterdam.
Watson DG (2015) *Pharmaceutical Analysis E-book: a textbook for pharmacy students and pharmaceutical chemists*. Elsevier Health Sciences, Amsterdam.

Chapter 8

Chromatographic Methods of Analysis

Shahid Shah[1], Malik Saadullah[2,*], Saeed Ahmad Rao[3], Reem I Alsantali[4], Ghulam Abbas[5], Sana Mehmood[2], and Muhammad Rashad[2]

[1]Department of Pharmacy Practice, Government College University, Faisalabad, Pakistan
[2]Department of Pharmaceutical Chemistry, Government College University, Faisalabad, Pakistan
[3]Faculty of Pharmacy, Islamia University of Bahawalpur, Pakistan
[4]Department of Pharmaceutical Chemistry, College of Pharmacy Taif University, Saudi Arabia
[5]Department of Pharmaceutics, Government College University, Faisalabad, Pakistan

Abstract

For the separation, purification, identification and characterization of the various components, separation methods are employed. The most viable tools for the separation and identification, e.g., proteins, monomers, polysaccharides, nucleic acids etc., is the chromatography employed by analytical chemists, biologists and biotechnologists. In the growth of many biological disciplines, for example, biotechnology, this method played an essential role. Chromatography can include one stage precipitation method with advance devices in several stages. It is extremely crucial for the protein purification utilized in biological labs. There are three primary forms of chromatography including gas

*Corresponding Author's Email: Maliksaadullah@gcuf.edu.pk.

In: Functional Analytical Techniques in Pharmaceutical Chemistry
Editor: Malik Saadullah
ISBN: 979-8-88697-059-3

chromatography, fluid chromatography and chromatography of supercritical liquids.

Keywords: chromatography, stationary phase, mobile phase, separation technique, proteins

Introduction

The components of chromatography mixture are used for qualitative and quantitative examination. It is a very significant approach and it enables the components of a mixture to be purified and identified. To identify the variations of size, load, binding affinities and other features present in a blend for the separation of the various components, extensive variety of chromatographic techniques are utilized. It is a great approach for isolating science from each discipline. It is a biophysical approach used to separate components from a mixture based on structure, size and other biochemical characteristics.

History of Chromatography

In 1903, by employing chromatography, a Russian botanist M. Tswett removed the plant pigments from the leaf. He used a solution through a solid absorbent column. The principle of component separation may be either partition or adsorption. The stationary phase is fluent if the principle is partitioning and the stationary phase should be solid in adsorption. In the mobile phase, liquid or gas is utilized. Drop Chromatography is employed in 1938 by two Russian scientists Izmailov and Shraiber, to separate components. It was later developed in thin layer chromatography by American scientist Kirchner.

Classification of Chromatography

Chromatographic techniques can be categorized according to the kind of balance procedure.

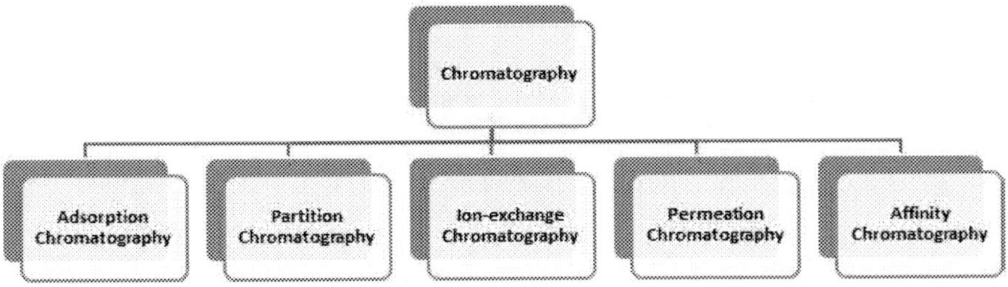

Adsorption Chromatography

Adsorption chromatography is a form of liquid chromatography. Due to their desorption and adsorption in front of the support, chemical substances are preserved in adsorption chromatography and function as a stationary phase. Chromatography adsorption is also sometimes referred to as solid liquid tubing. In adsorption, a chemical retention chromatography is based on the binding power of chemical for support, the supporting surface area, the moving phase quantity displaced from the chemical support and the binding resistance of the mobile phase to the support. The retention of the adsorption chromatography may be changed by electrostatic interactions, dipole-dipole, hydrogen bonding and dispersive interactions.

The binding force of the mobile phase with support is explained in adsorption chromatography by the eluotropic power of the mobile phase. A solution or liquid with a great eluotropic power to sustain a recognized item is a strong mobile phase. Since this mobile phase tends to bond to the support strongly and to make the analyst elute more quickly and spend more time on the mobile phase. For example, a methanol which is a polar solvent is more likely than a carbon tetrachloride which is a non-polar solution to have more elutropic power for a polar support like silica.

In adsorption chromatography generally three types of adsorbents are used:

1. Polar acidic supports
2. Polar basic supports
3. Non-polar supports

Silica is the most often utilized support for acid and polar chromatography adsorption. On this support the surface silanol groups have tendency to adsorb polar chemicals and to work for fundamental substances as well. Alumina is the principal form of polar and basic adsorbent in adsorption chromatography.

Alumina retains polar chemicals such as silica, although alumina works well for polar acid substances. Florisil is another basic and polar medium which can be employed instead of alumina, for example when a catalyst degradation with finished material is detected. The two forms of supports that may be utilized as include charcoal and polystyrene.

Partition Chromatography

Partition chromatography is a technique of separation. The components present in the mixture are separated into two liquid phases in partition chromatography because of differences in partition coefficients. The molecules are divided between two phases i.e., both stationary phase and mobile phase are in same phase in partition chromatography. As a result, molecules get isolated into either phase preferentially. Many separation processes like high-performance liquid chromatography, paper chromatography and high-performance thin layer chromatography and gas chromatography works on the principle of partition chromatography.

Ion-Exchange Chromatography

The chromatography involving ion exchange is based upon the opposing charged interaction between stationary phases. This is combined with the functional group that is positively or negatively charged. This approach is economical. It is a fast process of separation. It is used for hard water softening. This method also leads to demineralization. The chromatography of the ion exchange is the separation technique for loaded molecules by contact with the stationary opposite load phase in the form of ion resin.

Permeation Chromatography

Polymeric material containing different molecular pores is utilized as a stationary phase in permeation chromatography. Analysts as solved molecules in the mobile phase have their capacity to enter a sieve-like composition, depending on their size. The bigger molecules that are not integrated into the pores are maintained in the mobile phase. This technique is best suited to separate mixtures with a considerable difference of molecular dimension in the solutes. In this method, liquid or gas may be utilized as a mobile phase. Chromatography used for the separation of organic molecules and purification of organic synthetic polymers is called size exclusion chromatography.

Affinity Chromatography

The most potent method utilized for the separation of macromolecules such as nuclear acid and proteins is a form of chromatography. This method is also utilized for enzyme and antibody separation. It is a kind of separation process in which a mixture consists of its affinity to the system stationary stage separating its components. The widely used mobile phase of chromatic technologies consists of a certain amount of solvents as acetic acid, water, acetone, alcohol or more.

Chromatographic Theories

Mainly two theories are applied in chromatography.

Plate Theory

The equilibrium from the mobile phase to the stationary phase can be defined. The theory may also determine the volume and effectiveness of the columns.

Martin & Synge introduced Plate Theory. In this theory, a number of neighbouring imaginary segments termed theoretical plates (N) is split. The particles are fully balanced in fixed and portable phases under each theoretical platform. With the increasing range of theoretical plates (N), the plate length (H or HETP) decreases and the isolation is enhanced.

$$L = N \times H \text{ or } H = L/N$$

Where L is the column length, N is the number of plates, and H is the plate height.

Rate Theory

In this theory, the mobility of molecules in a column is determined. This might be the form of the band, its extension and its dispersal.

A more realistic depiction of the processes in a column takes the time needed to balance the solution between the stationary and moving phases (unlike the plate model, which accepts equilibration is substantially debauched). Consequently, eluent intensity affects this pixel band shape in the picture. The different methods in which individual particles pass through gel filter units are often impacted. The Van Deemter equation is derived from many reasons which contribute to the expansion of the band.

$$HETP = A + B / u + C u$$

where u is the average velocity of the mobile phase. A, B, and C are factors which contribute to band broadening.

A - Eddy diffusion

The mobile phase passes through the stationary phase of the column. Solvent particles simultaneously move in different directions during the stationary phase. Since the paths differ in length, the solid phase group might be expanded.

B - Longitudinal diffusion

The amount of analyte near the borders of the band is lower than in the centre. The sample solution absorbs from the centre to the edges. This results in an expansion of the band. If the mobile phase velocity is strong, this tends to spend less time on the columns and decreases the spatial propagation effect.

C - Resistance to mass transfer

It takes time for the analyst to find an equilibrium between the organic and aqueous phases. The analyte in the mobile phase can move ahead with the stationary phase analyte if the variable stage speeder is high and solvents have excellent preference for the solid surface. The band of analysts was expanded. The degradation is commensurate with the speed of the working step.

Van Deemter Plots

A plate height compared to the average linear moving speed.

When a solution is applied to an extractor solvent and passes through a firm phase (most of which is a solid surface), chromatography is theoretically based on certain components of the mixture are linked to the stationary phase. The rest is moved concurrently, including the mobile phase. Here are two essential modules of all methods of chromatography.

Paper Chromatography

The sort of separation process is paper chromatography. For separation in paper chromatography a specific paper is utilized.

Principle of Paper Chromatography

On two separate principles there is a double kind of paper chromatography. The first is chromatography with paper adsorption. The chromatography for paper adsorption is based on the changing degree of contact between the stationary and mobile phase. The molecules with a larger affinity stay adsorbed throughout the column for a longer duration. In contrast, the weaker affinity molecules travel quicker, allowing for the separation of molecules in various fractions. The second form of paper chromatography is chromatography using paper partitions. Moisture on the cellulose paper is the stationary phase for moving molecules with the mobile phase and this is the principle for chromatography of paper partitions. The molecular partition consequently depends on the strength of its adsorption in the stationary phase. A further notion of retention factor is used in chromatography on paper during the separation of molecules. The retention value of a molecule is determined in relation to the distance of the molecule from the mobile phase. Different compounds can be utilized to identify these molecules using retention values.

Method of Paper Chromatography

The stationary phase is a high-quality cellulosic paper. Different mixtures are employed as moving phases for inorganic and organic solvents. The sample solution is injected at the base of the paper approximately 2 μl to 200 μl and is allowed to dry in air. Thsample-filleded paper is then introduced into the mobile phase gently. The height should not exceed 1 cm. After the mobile phase is close to the paper border, the paper is removed. Determining the retention factor and detecting the segregated components using different approaches. Figure 8.1 describes an example of paper chromatography.

Applications of Paper Chromatography

Paper chromatography is carried out to test the purity of various medicinal goods. In diverse samples such as drinks and food, contamination can also be detected. This approach may also be used to separate contaminants from other industrial goods. Reaction mixtures are sometimes examined with paper chromatography in chemical laboratories.

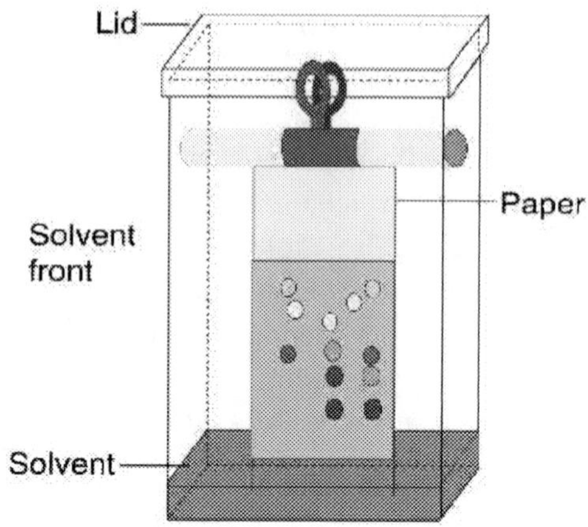

Figure 8.1. An illustration of paper chromatography.

Thin Layer Chromatography (TLC)

The thin-layer chromatography is a kind of technology of separation employed as a fixed phase with a liquid mobile phase for a thin layer on a solid supporting sheet.

Principle of Thin Layer Chromatography

Thin layer chromatography is based on the premises that the components that have a fine connection to the stationary phase are then separated from the components of a mixture. Instead, the mobile phase elutes other components. The ligand/substratum is linked to a fixed phase in order to reveal the reactive areas for connecting components. Elements that have binding sites for the substrate at the stationary phase bind the mixture while the other components are eluted with the mobile phase. It also passes through the mobile phase. The molecules are seen as dots throughout the stationary phase following separation at a different place. Different approaches for detecting molecules are employed.

Method of Thin Layer Chromatography

As a standard phase of solid support, for example, thin platform, glass, or aluminium foil can be utilized and dried. The sample is injected at the stationary phase approximately 1 cm above the border of the plate. The sample-loaded plate then came into the mobile phase cautiously. It must not exceed 1 cm in height. When the moving step approaches the edge of the platform, the plate is removed. It is computed in paper chromatography as its retention factor. Different methods identify the distinct components. A thin-layer chromatography example is shown in Figure 8.2.

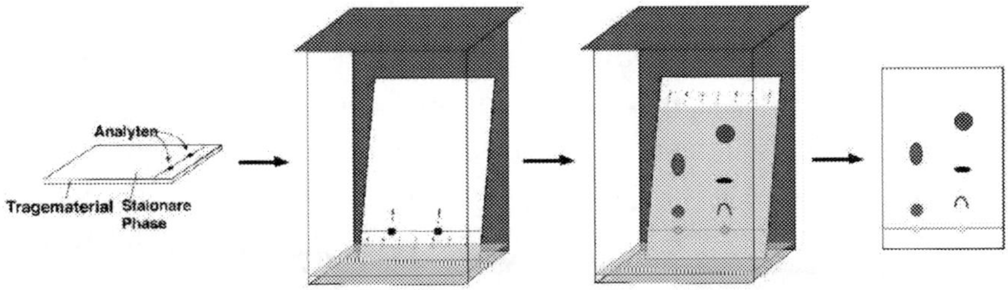

Figure 8.2. An illustration of thin layer chromatography.

Applications of Thin Layer Chromatography (TLC)

In laboratories, several chemicals contained in a mixture are identified using thin-layer chromatography. This approach is used to analyze fibres in forensic. Thin chromatography layers also permit the assessment of different pharmaceuticals. This method is also used to compose and identify medicinal herbs.

Column Chromatography

In column chromatography, a fine adsorbent material is mainly employed as the steady stage. In chromatography, the column generally seems to be similar to the pipette of Pasteur. This method is used to purify, characterize and isolate active components in mixed samples. This method may be utilized with a wider spectrum of mobile phases. However, this procedure takes time. The column chromatography is a method used for the separating mixture of components. Based on their difference adsorption in the fixed stage, these components are separated. This means that while passing through a column, they might move at different rates. It is a method of solid-liquid chromatography employed as a stationary phase by the solid and by mobile phases by the liquid or gas.

Principle of Column Chromatography

The idea of differential adsorption is based on this approach when various molecules in a mixture have varying affinities towards the absorbent present at the stationary stage. The molecules of high affinity remain in the stationary phase and decrease their movement speed across the column for a longer period. On the other hand, the compounds with a low composition travel faster so that they can be split into pieces. This is also called an absorbent phase of a stationary (usually silica) phase in the column chromatography and the mobile phase is a liquid that permits the molecules to pass through the column smoothly. An illustration of column chromatography is described in Figure 8.3.

Method of Column Chromatography

A glass tube is used as column. The tube of this glass has a thin and uniform cellulose or silica layer. By adding the mixture to the mobile phase, the sample is produced. The sample will be injected from the top into the column. The sample is passed here under gravity influence. The elution technique separates the molecules attached to the column, using either a similar polar solution or a different polarity sample. (Isocratic technique) and (gradient technique). The separated molecules can further be analyzed for various purposes.

Applications of Column Chromatography

For the isolation of contaminants from different biological blends, column chromatography is employed. It is also used for biological mixture purification. This method is utilized for the separation of active molecules and metabolites from various materials. The method for the identification of medication in crude extracts is increasingly employed.

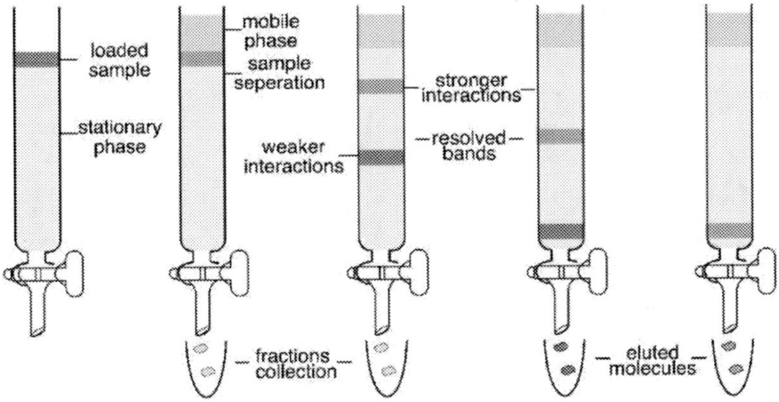

Figure 8.3. An illustration of column chromatography.

Gas Chromatography

This technology allows components to be vaporized without denaturation for the separation and purification of constituents. The components of the mixture are dissolved using a solvent in this procedure. The components are vaporized in order to separate analyte by the distribution of mixture between two phases. Phases can be classified as either fixed or movable. An inert gas is utilized as a mobile phase in order to transport molecules from the sample to the heated column during analysis.

There is just one form of chromatography that does not need a mobile phase to interact with the sample and that is gas chromatography (analyte). In GSC and GLC, the stationary phase can be either a solid adsorbant or a liquid on an inert support. Using gas chromatography, molecules are sorted depending on their retention period. Because of the molecular affinity for the stationary phase, retention time is affected.

Principle of Gas Chromatography

Greater affinity components will have a longer retention period in gas chromatography, while components with lower affinity will have a quicker exit from the column. Because of their increased affinity for stationary phase while travelling through mobile phase, the components on the other hand will have less retention duration. Helium gas is the most common mobile phase. Helium carries the sample up and down the column as it passes through.

Injectable vapour is created from the sample. A detector is then used to determine the retention time. They are gathered individually because they exit the stationary phase at various periods.

Method of Gas Chromatography

It is injected into the column after it has been vaporized. Samples are vaporized and then mixed with the mobile phase, which is then pumped through the rest of the column. As the molecules are separated based on their affinity to the stationary phase, the column is filled with a stationary phase. The components of the mixture arrive to the detector at various times due to their varying retention periods in the column. Below is an example of gas chromatography Figure 8.4.

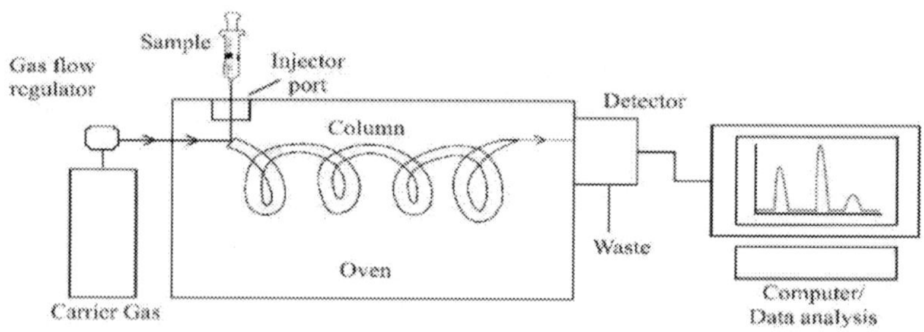

Figure 8.4. An illustration of gas chromatography.

Instrumentation of Gas Chromatography

Gas chromatography consists of following parts.

Sample Injection

Adding the sample mixture at the top of the column requires a port. Heated sample ports are commonly used in modern injection techniques. These ports allow synchronized sampling and vapourization. When inserting samples through a rubber barrier, the microsyringe is utilized to inject them into the vaporization compartment. Only a few microliters of sample can be used. To begin the separation, only a tiny percentage of the sample volume is needed. A sample splitter is used to send surplus sample to the garbage stream. Commercial gas chromatographs can employ split and splitless injections.

Temperatures in the vaporization chamber are typically kept at 50°C. As a result of mixing the sample combination with carrier gas at this temperature, the sample mixture may be transferred into the column at a higher temperature than its boiling point.

Carrier Gas

Different types of carrier gases can be employed in gas chromatography, and they play a crucial role in this method. It should be devoid of oxygen and completely dehydrated. Gas chromatography should utilize a chemically inert mobile phase. A lot of gas chromatography experiments employ helium as the mobile phase since it is a safer technique alternative to hydrogen. It is also more efficient technique and has higher flow rates. In addition, it is compatible with a wide range of detectors. Other gases such as hydrogen, nitrogen and argon can also be utilized, depending on the detector performance requirements and intended application. Due to greater flow rates and lower molecular weight, hydrogen and helium are commonly employed in traditional detectors such as flame ionization and thermal conductivity. This results in shorter research times and lower elution temperatures. Thermal conductivity detectors work well with hydrogen or helium as a carrier gas. This is due to the fact that helium/hydrogen and organic vapor have a larger thermal conductivity differential than other carrier gases. Using nitrogen or argon gases, mass spectroscopy is another detector. Their molecular weights are greater and the vacuum pump's efficiency has increased. Gas carriers are stored in pressurized tanks. Controlling the flow of gas is done via pressure regulators, flow metres and gauges. The gas should have a purity range of higher than 99 percent and a low oxygen with hydrocarbon content. This amount should be less than 0.5 ppm, according to the World Health Organization. This system has a molecular filter for water removal. Depending on the gas, several types of regulators are employed.

Column Oven

In order to carry out the experiment with precision, ovens are used to regulate the temperature of the column. Ovens may be programmed in two ways: isothermal and temperature. In isothermal programming, the optimal column temperature is the midway point of the sample's boiling range. It will be easier to use isothermal programming if the sample boiling point range is limited. Wide band expansion is sluggish to elute the fraction with high boiling points when the column temperature is kept low under isothermal programming. Elevating the temperature till it reaches the boiling point for the higher boiling components resulting in abrupt peak elution. If the temperature is lowered, there will be no separation since the elutes will flow rapidly.

In temperature programming, the temperature of the column is raised in stages or gradually during the separation process. If you want to separate mixtures with a broad variety of boiling points. When the procedure is begun at a low temperature, low boiling components are identified. In order to identify the less volatile and high-boiling components, the temperature is increased during the separation of the sample.

Types of Column

The column of gas chromatography may be of following types.

Open Tubular Columns and Packed Columns

Open tubular columns are also called capillary columns. They are present in two fundamental types.

Wall-Coated Open-tubular (WCOT) Column

As the name suggests, they are capillary tubes with columns coated with a thin coating of the stationary phase on their column walls. A small coating of adsorbant solid, similar to single-celled sea-plant skeletons, is initially applied to the column walls before the column is filled with water. This is followed by the employment of the liquid stationary phase to treat adsorbant solid. There is an increase in the efficiency of the WCOT columns. The WCOT column is mostly composed of glass. WCOT column may be made from copper, stainless steel, aluminium, and plastic. When utilized, each material has its own advantages. Gaseous and concentrated HCl can be treated using WCOT columns made of glass.

Support-Coated Open-tubular (SCOT) Column

They can hold more stationary phase than WCOT columns. Their sample capacity is much higher.

Fused Silica Wall Coated (FSWC) Column

It is a WCOT column of a particular sort. FSWC is made from silica that has been purified to remove any metal oxides. Comparatively, FSWC columns are significantly thinner, having 0.1mm diameters and are 100m long as opposed to glass columns which are substantially thicker.

Applications of Columns

The kind of sample determines whether or not a column is used. Example: FSWC column is used to analyze blood alcohol levels in a laboratory setting. As a result, this column is referred to as Zebron-BAC. As with Zebron-inferno, it is utilized for high-temperature processes. Hydrocarbons are separated and purified using this process. This column may also be used to separate acidic and basic materials.

Detection Systems

Different types of detectors are used in gas chromatography.

1. Mass Spectrometry Detectors
2. Flame Ionization Detectors
3. Thermal Conductivity Detectors
4. Chromatogram
5. Electron-capture Detectors
6. Atomic Emission Detectors
7. GC Chemiluminescence Detectors
8. Photoionization Detectors

Applications of Gas Chromatography

Using gas chromatography, volatile mixtures are physically separated. In addition to cosmetics, medicines and environmental sciences, it may be employed in a wide range of disciplines. Other applications include the study of human breath, blood, saliva and other body fluids in the medical sciences.

Applications of Gas Chromatography in Food Analysis

In order to determine the quality and quantity of food, gas chromatography is employed. In gas chromatography, food components, flavor, fragrance, and additives are assessed. Gas chromatography is also used to study contaminants in food products such as pesticides, fumigants, pollutants, and poisons.

Pharmaceutical Applications of Gas Chromatography

There are several ways to separate polar molecules from one another and analyze them using gas chromatography. Using this method, volatile chemicals may be separated. In future, these chemicals will be utilized to create medicinal agents.

Applications of Gas Chromatography for the Analysis of Breath

Exhaled breath from smokers and passive smokers can be analyzed using gas chromatography (GC). Gas chromatography-mass spectrometry was used in this study (GC-MS). Smokers' and passive smokers' samples were analyzed for a variety of substances. There are a number of compounds that may be found in the environment such as dimethylfuran, furan, 2-butanone, octane, acetonitrile, decane, and benzene.

Applications of Gas Chromatography in Fatal Strychnine Poisoning

As part of the historical and toxicological study of suicidal instances that resulted from lethal strychnine poisoning, gas chromatography is utilized. In order to characterize and identify strychnine in body fluids and tissues, gas chromatography was employed. In addition to subclavian blood, inferior vena cava blood, urine, and bile can be tested for strychnine concentrations.

Application of Gas Chromatography in Evaluation of Human Body Fluids

Gas chromatography is used to evaluate human body fluids in healthy and sick conditions. In this method, different organic acids are utilized.

Application of Gas Chromatography for the Study of Cardiovascular Disease

A number of isoprostanes have been found in tissues and body fluids by employing GC-MS techniques. This group of isoprostanes is a key catalyst for lipid peroxidation, which can lead to a variety of illnesses, including cardiovascular disease and cancer.

Ion Exchange or Ion Chromatography

On the basis of the interaction between stationary phases, ion exchange chromatography operates. It is important to note that these stationary phases have opposing charges. This is paired with a functional group that is positively or negatively charged. This is a cost-effective method. This is a fast way to separate things. Softening hard water is a common application for this product. As a result of this method, demineralization is also achieved. When charged molecules interact with an oppositely charged stationary phase such as ion-exchange resin separated by using the ion exchange method.

Principle of Ion Exchange Chromatography

As the name implies, this method relies on the interaction between the analyte and the resin. For the elimination of the charged molecules, the ions are exchanged. Various charges are employed to coat a stationary phase to which oppositely charged particles of the mixture will adhere. Resin exchanges ions (cations/anion) with greater affinity for charged components. After that, bind with components, displace the oppositely charged resin and then assemble the final product. To remove the cation or anion exchange resin component complex, specific buffers must be used to remove the complex.

Method of Ion Exchange Chromatography

As stationary phase, a positively or negatively charged resin is introduced into the column. In a column, charged particles bind to resins that have opposing charges. Cation-exchange resin repels negatively charged particles by binding to positively charged ions. An anion exchange resin, on the other hand, repels positively charged resin because the negatively charged particles bond to it. After that, the column is treated with an appropriate buffer that separates the charged exchange resins from the charged particles. Schematic diagram of the process of ion-exchange chromatography is described in Figure 8.5.

Applications of Ion Exchange Chromatography

Ion exchange chromatography is used to purify water by replacing cations with hydrogen ions and anions with hydroxyl ions. Nucleic acid hydrolysis is also analyzed using this method. Separating metals and other inorganic components is another application of this technique.

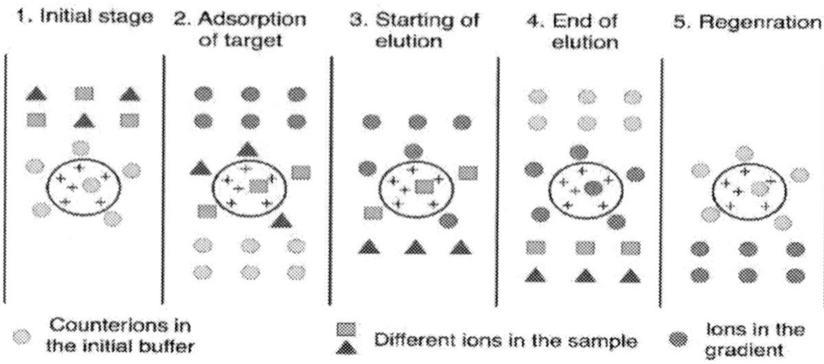

Figure 8.5. Schematic diagram of the process of ion-exchange chromatography.

Anion Exchange Chromatography

Using iion-exchange resin and anions (negatively charged molecules), anion exchange chromatography separates anions (negatively charged molecules) from the cations (positively charged molecules).

Principle of Anion Exchange Chromatography

A positively charged resin and a negatively charged analyte interact in this method. Here, the positively charged ions displace the negatively charged components in order to remove them from the system. To coat the stationary phase, particles of the mixture with negative charges will attach and positive charges are applied to the surface. Resin exchanges ions (cations/anion) with greater affinity for charged components. After that, they bond the components, displace the oppositely charged resin and then assemble the final product. To remove the cation or anion exchange resin component complex, specific buffers must be used to remove the complex.

Method of Anion Exchange Chromatography

As the stationary phase, a positively charged resin is added to the column. In a column, charged particles bind to resins that have opposing charges. Cation-exchange resin repels negatively charged particles by binding to positively charged ions. An anion exchange resin, on the other hand, repels positively charged resin because the negatively charged particles bind to it. After that, the column is treated with an appropriate buffer, which is used to separate out the charged exchange resins from the charged particles. Process of anion exchange chromatography is described in Figure 8.6.

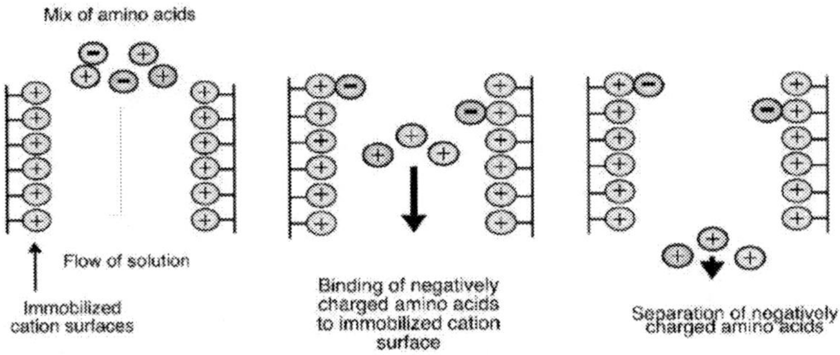

Figure 8.6. Process of anion exchange chromatography.

Applications of Anion Exchange Chromatography

Proteins and amino acids can be separated from their mixtures using anion exchange chromatography (ACEC). Negatively charged nucleic acids are separated using this method and are utilized for additional nucleic acid analysis. This method replaces anions with hydroxyl ions to purify water. Anion exchange resins are used to separate metals that have negatively charged complexes.

Cation Exchange Chromatography

Anions (negatively charged molecules) interact with cations (positively charged molecules) in the form of ion exchange resin to separate them.

Principle of Cation Exchange Chromatography

As the stationary phase, a negatively charged resin is introduced into the column. It is then transported down the column where the charged particles bind to resins that are charged in the opposite direction to the particles.

Method of Cation Exchange Chromatography

As the stationary phase, a negatively charged resin is introduced into the column. In a column, charged particles bind to resins that have opposing charges. Cation-exchange resin repels negatively charged particles by binding to positively charged ions. An anion exchange resin, on the other hand, repels positively charged resin because the negatively charged particles bind to it. This is then followed by the use of an appropriate buffer in order to separate out both the complex of charged exchange resins and their charged particles from each other. Process of cationic exchange chromatography is described in Figure 8.7.

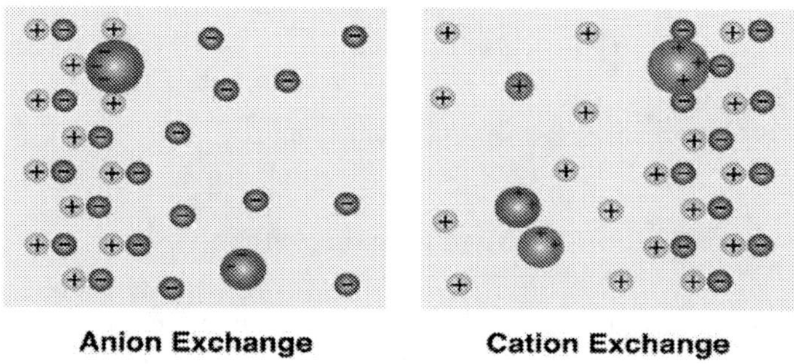

Figure 8.7. Process of cationic exchange and anionic exchange chromatography.

Applications of Cation Exchange Chromatography

Analysis of components following nucleic acid hydrolysis is done by using Cation Exchange Chromatography (CEC). Similarly, the metal ions themselves bind to the negatively charged resins to remove the negatively charged complexes in this method of separation. Using cation exchange chromatography, hydrogen ions are exchanged for positively charged ions in order to purify water. Rocks and other inorganic substances can also be analyzed using this method.

Size Exclusion Chromatography

By using size exclusion chromatography, the molecules are separated according to their sizes. Gel filtration or gel permeation chromatography are other names for it. An extremely little amount of sample must be used to carry out this procedure successfully. There is a significant reduction in the time to separate compounds. Proteolytic fractionation and purification of proteins, carbohydrate chains and peptides are some of the applications of this technique. Description of size exclusion chromatography is depicted in Figure 8.8.

Instrumentation of Size Exclusion Chromatography

Parts are moved at different rates and then analyzed in discrete intervals from the stack in the SEC process. During capillary electrophoresis or gel electrophoresis, bulk solutions are frequently held back by the top of the column packing, the stationary phase, or a combination of both (stationary and mobile phase). While molecular size or molecular hydrodynamic volume influences the separation process of SEC and is unaffected by the type of mobile phase, unlike gas chromatography (GC) and liquid chromatography (LC). Because of its great affinity for the stationary phase and good solubility in samples, the mobile phase characteristics are still significant. Make sure that the polymer components are properly suspended in liquid by thoroughly dissolving a substance. Accordingly, the solutes are mixed as they move through a porous column. There are three main ideas in the SEC's fundamental configuration: the stationary phase (column), mobile phase (solvent), and sample preparation.

Applications of Size Exclusion Chromatography

With remarkable success, size exclusion chromatography has been used to separate proteins and polypeptides, viruses and enzymes, antibodies and hormones, nucleic acids and polysaccharides as well as a variety of liquids such as asphalts, polyethylenes, butyl rubbers, silicon polymers and polystyrene. Separating big molecules of biological origin from inorganic and ionizable species is a major use of this technology Desalting is the process of removing the salt from the food. If you want to separate haemoglobin from sodium chloride, you may do it with a Sephadex gel column. Preparation is also possible using this approach.

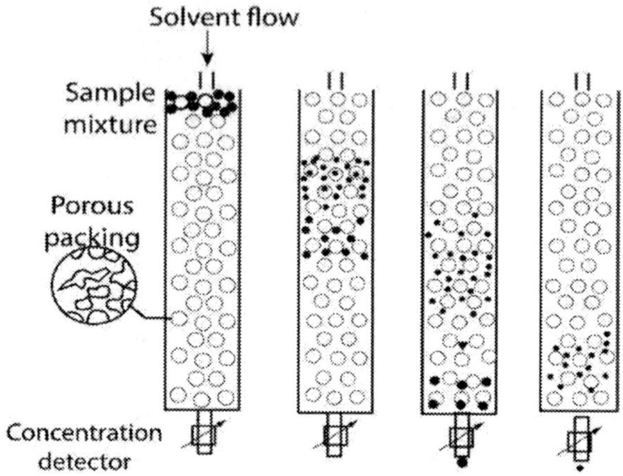

Figure 8.8. Process of Size Exclusion Chromatography.

High Performance Liquid Chromatography (HPLC)

As an example, a column chromatography that is changed a separation of components in a mixture based on their affinity for the stationary phase in high-performance liquid chromatography.

Principle of High-Performance Liquid Chromatography

On the basis of differential adsorption, various components in a mixture interact with the absorbent on the stationary phase in varying degrees. This slows down the movement of the molecules with a higher affinity. Although faster-moving and lower-affinity molecules can be segregated into different fractions because of their speed. Due to the high pressures (up to 400 atmospheres), this method is similar to column chromatography in that case the solvent must be forced instead of dripping. Schematic representation of HPLC is shown in Figure 8.9.

Method of High-Performance Liquid Chromatography

To prepare the column, a dry glass tube is utilized. Thin coating of cellulose or silica is applied to this column (stationary phase). The sample is then produced by adding the mixture to the mobile phase. Sample is injected into the column from the top and a high pressure pump is employed to keep the sample flowing at a consistent pace. It then goes to a detector that identify molecules at a certain wavelength of absorption. Separated molecules can be utilized for a variety of reasons.

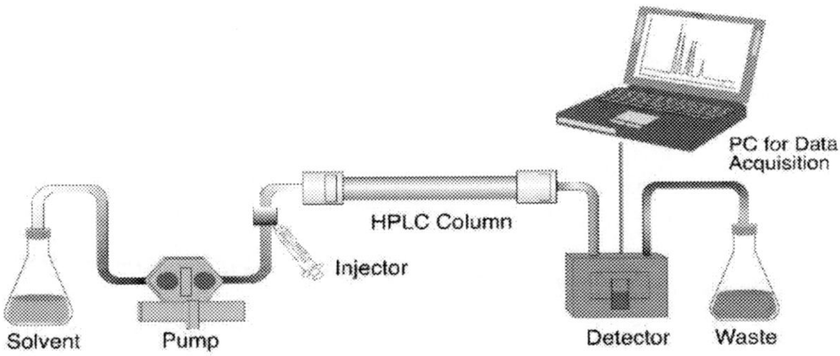

Figure 8.9. A schematic diagram of HPLC.

Applications of High-Performance Liquid Chromatography

High performance liquid chromatography is used to analyze environmental samples for contaminants. Purity and quality control of diverse industrial productions are maintained by using this technology. This method is also used to separate other biological substances such as proteins and nucleic acids.

Fast Protein Liquid Chromatography

As a kind of liquid chromatography, fast protein liquid chromatography (FPLC) may be used to separate and characterize proteins from mixtures as well as to purify other substances such as oligonucleotides or plasmids. It is also known as fast performance liquid chromatography. For improving biotechnology, biochemistry, and pharmaceutical chemistry, new methods for studying and purifying proteins must be developed. For the creation of novel medications and treatments to treat various ailments, the development of new techniques and procedures would be beneficial.

FPLC is a method for protein separation that is based on the idea of adsorption. The protein is isolated from the other elements and components in the sample combination before it is submitted to the laboratory. Aqueous buffer is utilized as the mobile phase, while agarose beads are employed as the stationary phase. In various reservoirs, the composition of the buffer may differ. Protein Liquid Chromatography is a popular technique in the fields of proteomics, enzymes, and biology. To identify the presence of particular proteins that are responsible for various illnesses, this test is utilized.

Principle

In rapid protein liquid chromatography, a microprocessor is utilized to maintain a consistent flow rate of solvents using a software interface. The mobile phase consists of an aqueous buffer and the flow velocity is maintained by a positive displacement pump while the buffer composition may be changed by pulling fluids from external reservoirs (see figure 8.10). As the stationary phase, cross-linked agarose beads are often employed. These beads were arranged in a cylinder shape. After passing through a detector, the eluent is analyzed to determine the protein content and salt concentration.

Method of Fast Protein Liquid Chromatography

The fast protein liquid chromatography system consists of following components:

1. control unit
2. high-precision pumps
3. column, detection system
4. fraction collector

As a result, the proportion of each buffer entering the column may be continuously varied with the pump. It is filled with the sample solution that will be injected into the column by an injection loop with a tubing segment of known volume. The injection valve links the mixer and sample loop to the column that loads the samples. Columns are made from either glass or flexible plastic and filled with tiny droplets of gum arabic (mastic). As the buffer flows downhill from the top to bottom, it is placed vertically. Cells are used to test protein content in eluent. The detector records the salt and protein concentrations at the end. Process of fast protein liquid chromatography is described below in Figure 8.10.

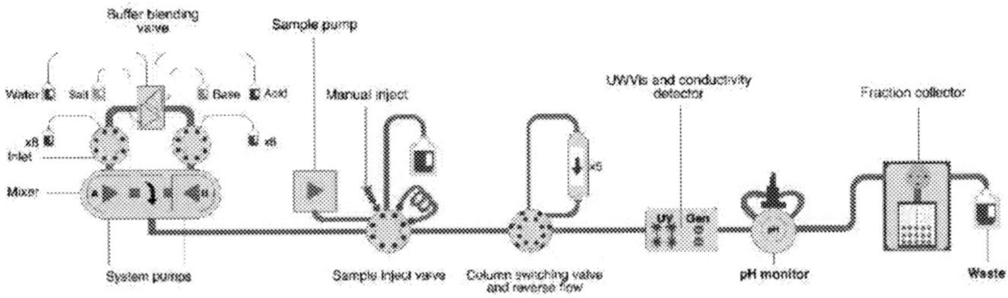

Figure 8.10. Process of fast protein liquid chromatography.

Applications

Diagnosis of β-thalassemia

β-thalassemia is a hereditary blood disease in which haemoglobin production is reduced. *β*-thalassemia is diagnosed by measuring HbA2 (haemoglobin A2) levels quantitatively. Traditional methods of testing for *β*-thalassemia include high performance liquid chromatography (HPLC), electrophoresis and microcolumn chromatography. Hemoglobin A2 levels are now determined using the fast protein liquid chromatography (FPLC) technique. Compared to previous procedures, the technique is simple, quick, repeatable and less expensive.

Separation of Plasma Proteins in Urine and Cerebrospinal Fluid

Quick and repeatable characterization of cerebrospinal fluid and urine plasma proteins is possible with anion-exchange liquid chromatography. The elution properties of 21 plasma proteins were determined using fast protein liquid chromatography (FPLC). Fast protein liquid chromatography was used to profile urine proteins, tubular proteinuria, and hemoglobin-derived products.

Ultra Performance Liquid Chromatography

As an alternative to conventional liquid chromatography, Ultra Performance Liquid Chromatography (UPLC) is a novel and eco-friendly chromatographic technology. There are two major differences between it and traditional liquid chromatography: the average particle size is 1.7 microns, as opposed to 5.0 microns for typical columns and the system operates at 15,000 pressure instead of the customary 5000 psi. Unlike ordinary liquid chromatography, ultra-performance liquid chromatography equipment is design to withstand greater back-pressures than traditional Liquid Chromatography. It is possible to achieve greater speed, sensitivity and resolution with ultra-performance liquid chromatography. An illustration of UPLC is described in Figure 8.11.

So far, sample injection is a challenge in ultra-performance liquid chromatography due to the difficulty in building a valve that meets the requirements of sealing at high pressure while accurately injecting a very tiny volume of sample. We've come up with a variety of strategies.

1. For quantitative purposes, the static split injection method utilization is too unpredictable.
2. Technique of pressure-balanced injection
3. Injection time is decreased, repeatability is improved and sample waste is minimized. This approach is really simple to implement.
4. System for air-activated needle valve injection.
5. Modified diamond-coated six-port computerized switching valve with different-sized sample loops;

In the early days of capillary ultra-performance liquid chromatography, the main stationary phases employed were nonporous silica particles with a size of 1-1.5 m. According to the theory, nonporous particles reduce band broadening due to sample diffusion into the inert mobile phase within the pores, resulting in an overall quicker mass transport process.

Principle

According to Van Deemeter's equation, HETP or column efficiency is directly proportional to flow rate in ultra-performance liquid chromatography.

Instrumentation

The instrumentation of ultra-performance liquid chromatography includes:

1. Sample injection
2. Ultra performance liquid chromatography columns
3. Detectors

Sample Injection

The injector is used to properly measure a tiny volume of fluid and then inject it. The sample is present in the mobile phase of this solution. As a result, the injection must be accurately and consistently injected. Because traditional injection valves cause significant pressure instabilities, ultra-performance liquid chromatography requires a pulse-free injection method. Because of the possibility for band spreading, the swept volume of the device should be kept to a minimum. In order to take full use of ultra-performance liquid chromatography speed, a quick injection set time is required. There should be smaller volume injections and minimum carryover necessary to increase the sensitivity. The volume of the sample in ultra-performance liquid chromatography is typically 2-5 μl. For the examination of biological materials, direct injection method is being employed.

Ultra Performance Liquid Chromatography Column

Columns used for ultra-performance liquid chromatography have been designed and prepared by the following different companies:

1. Waters
2. Agilgent technology
3. Altech Assosciate.
4. Phenomenex

Various types of columns that are used in ultra-performance liquid chromatography are packed with components which are formed through different technologies. These technologies are given below:

1. CSH (Charged Surface Hybrid) particle technology
2. BEH (Ethylene Bridged Hybrid) particle technology
3. HSS (High Strength Silica) particle technology
4. PST (Peptide Separation Technology).

Detector

There should be a high sampling rate with limited peak access, and the peak distribution should be as small as possible to prevent waste of the separated solute. The separation sensitivity of ultra-performance liquid chromatography is two to three times higher than that of high performance liquid chromatography, the preceding analytical technology. UPLC uses PDA (Acquity Photodiode Array) and TUV detectors which utilize Teflon (AF) that creates an internally reflecting plane and enhances light transmission efficiency by reducing internal absorptions.

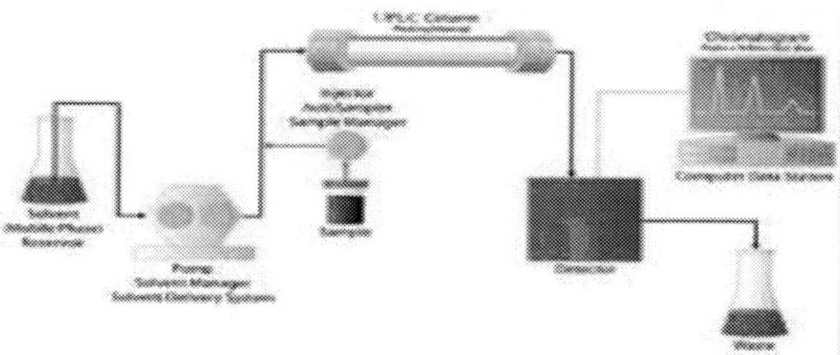

Figure 8.11. An illustration of UPLC.

Applications of Ultra Performance Liquid Chromatography

Determination of Pesticides in Groundwater

UPLC^{TM}-MS/MS (triple quadrupole tandem mass spectrometry) and ultra-performance liquid chromatography can be used to detect trace levels of pesticides in groundwater in a short amount of time. Analysis is more sensitive, faster and more precise due to ultra-high-performance liquid chromatography.

Improvement of Resolving Power in Peptide Maps

Peptide mapping is one of the most significant techniques for analyzing proteins. In comparison to high-performance liquid chromatography, ultra-performance liquid chromatography delivers considerably enhanced sensitivity, resolution, and peak capacity for tryptic digests of phosphorylase. A detailed description of the protein is provided.

Rapid Dose Formulation Analysis

Pharmaceutical applications are increasingly using ultra-performance liquid chromatography with UV and MS detection. To examine the efficiency of separations with changes in flow rate, a variety of commercial medication formulations were utilized as models. Resolution, retention duration, theoretical plates, column starkness and peak area were used to determine the efficiency. For example, the separation of chloramphenicol and mefenamic acid in a polyethylene glycol-200/dimethylacetamide vehicle was examined.

Analysis of TCM (Traditional Chinese Medicines)

Principal problems include measurement and identification of traditional Chinese medications using chromatography. Each component of traditional Chinese medicine plays an important part in its overall efficacy. As a result, the examination of all of the components is crucial for quality control at the same time. Using ultra-performance liquid chromatography, traditional Chinese medicines are tested for purity.

Multi-Residue Analysis of Pharmaceuticals in Waste Water

Different pharmaceutical firms use ultra-performance liquid chromatography in conjunction with Q-TOF-MS to verify and screen medicines (anti-ulcer agents, antibiotics and beta-blockers) found in samples of waste water treatment plant.

Identification and Characterization of Metabolites

A high sample throughput is required by the analysts for the identification and characterization of the metabolites that are critical to novel drug discovery in order to provide rapid findings for the medicinal chemists. For biomarker discovery, high performance liquid chromatography and mass spectrometry can be used together.

In Manufacturing/Quality Assurance (QA)/Quality Control (QC)

During the manufacture of a pharmaceutical dosage form and a drug product, essential factors such as identification, purification, quantification, efficacy, and protection must be examined. As part of quality assurance and quality control, material consistency is also examined. In quality assurance and quality control laboratories, ultra-performance liquid chromatography is used for quantitative and controlled analysis.

Method Development/Validation

Development and validation of methods is a time-consuming and difficult process. It is necessary to measure numerous combinations of parameters such as temperature, pH, mobile phase, gradient chemistry and column in order to build a robust and repeatable technique. This is an essential laboratory technique that reduces the cost and improves the efficiency of analysis both of which are crucial for the development and validation of a method. Separation speed and efficiency improved with ultra-performance liquid chromatography (UPLC) lead to the rapid development of new methods. The ultra-performance liquid chromatography column of high permanence allows for a wide variety of column pH and temperature settings.

Dissolution Testing

During the development and preparation phase, testing for drug dissolution is a crucial step. Each batch stability and standardization may be verified with the dissolution data. The separation sensitivity of new and powerful formulations must be enhanced. Precision and reliability are provided by the ultra-performance liquid chromatography method which automates online sample collection.

Bioanalysis Studies/Bioequivalence

For the quantification of medicines in biological samples, pharmacokinetic investigations needed bioequivalence tests. If you are comparing newly produced medicines to their original formulation, this is a crucial step. Ultra-performance liquid chromatography combined with mass spectrometry produces consistent and reliable results because of its sensitivity and selectivity.

Toxicity Studies

A toxicology issue produces a drop in the number of drug candidates, which results in a loss of revenue for the company. Inhibition/initiation of metabolizing enzymes and toxicity in the body are difficult to predict. Because of its great resolution, ultra-performance liquid chromatography allows for precise detection. Aside from that, its high sensitivity allows for the identification of peaks at low concentrations as well. These characteristics minimize the amount of time it takes to do an analysis as well as the likelihood of a sample analysis failure.

Iodinated Disinfection Byproducts

Iodinated disinfection byproducts (IDBPs) in drinking water have been characterized using gas chromatography and mass spectrometry. In contrast, ultra-high performance liquid chromatography combined with electrospray ionization-triple quadrupole mass spectrometer utilized to characterize iodinated disinfection byproducts in samples of chlorinated and chlorinated-ammoniated water Seventeen iodinated disinfection by-products structures were calculated in this procedure.

Therapeutic Drug Monitoring

Two β-lactamase inhibitors and seven -lactam antibiotics in human plasma were measured using ultra-performance liquid chromatography combined with mass spectrometry.

Analysis of Explosives

Using ultra-high-performance liquid chromatography, a variety of explosives may be analysed. This approach requires a little amount of sample and takes less time.

Determination of Phytochemicals

It is possible to identify and quantify procyanadines, oligomers and monomers, flavonoids and phenolic compounds using ultra-performance liquid chromatography, as well as coumarins, isoflavones and alkaloids using ultra-performance liquid chromatography.

Identification of Static and Kinetic Lipid Phenotype

Lipid acid content and endogenous biosynthesis are examined using high resolution ultra-performance liquid chromatography combined with mass spectrometry.

Analysis of Free Amino Acids (FAA) in Wines

UPLC can investigate the development of free amino acids (FAA) in less aged white wines. The 6-aminoquinolyl UPLC™ technique is a well-known method for amino acid analysis. 24 amino acids may be obtained in 23 minutes using this innovative method.

Identification of Metabolic Biomarkers

There are no diagnostics that can accurately distinguish between patients with epithelial ovarian cancer (EOC) and healthy persons. UPLC/QTOF/MS was used to evaluate plasma samples from patients with epithelial ovarian cancer and healthy persons.

Drug Abuse

It is possible to employ the UPLC-MS/MS technology as a fast and precise screening platform for the characterization and quantification of benzodiazepines and opioids in urine.

Conclusion

Various chromatography methods have been developed some of them include adsorption chromatography, partition chromatography, ion-exchange chromatography, permeation chromatography, affinity chromatography paper chromatography, thin-layer chromatography (TLC), column chromatography, gas chromatography, size exclusion chromatography, high-pressure liquid chromatography, fast protein liquid chromatography, ultra-performance liquid chromatography.

In chromatography, components of a mixture are separated by using a chromatographic apparatus. Mobile phase: The sample mixture is dispersed in a fluid (such as air or water) that transports it through a stationary phase: A column, capillary tube, plate or sheet that is fixed with a substance known as the stationary phase. There are differences in the affinity of the different mixture elements to the stationary phase. Discrimination between the mobile and stationary phases is used to separate mixture components.

Chromatography may be divided into two categories: preparative and analytic chromatography. Both techniques are used to separate the components of a mixture. Usually preparative chromatography is used for purification purpose and analytical chromatography is used for smaller quantity of materials for determination of existence of analyte in a mixture with their relative proportions.

Conflict of Interest

None Declare

Consent for Publication

None.

Acknowledgment

None Declare.

References

AKTA FPLC system manual. 2000. Amersham Pharmacia Biotech AB. Uppsala. Sweden.

Boer, H. *Vapour pHase Chromatography,* London, 1957.

Bogusław B., Agnieszka U., Tomasz L., Natalia D. and Anton A. 2008. Analysis of exhaled breath from smokers, passive smokers and non-smokers by solid-phase microextraction gas chromatography/mass spectrometry. *BioMed Central.* 01:1141.

Cooper, E., Turner, R., Johns, A. E., Lindblom, H. and Britton, J. V. 1983. Applications Of Fast Protein Liquid Chromatography TM In The Separation of Plasma Proteins in Urine And Cerebrospinal Fluid. *Clinical Chemistry.* 29:1635-1640.

Dimbat, M., Porter, P. E. and Stross, F. H. 1956. *Analytical Chemistry*. 28, 290.

Dogan, A., Ozkara, S., Sarı, M. M. and Uzun, L. 2012. Evaluation of human interferon adsorption performance of Cibacron Blue F3GA attached cryogels and interferon purification by using FPLC system. *Journal of Chromatography.*

Driscoll, J. N. Review of Photoionization Detection in Gas Chromatography: The first Decade. *Journal of Chromatographic Science,* Vol 23. November 1985.488-492.

Edzard S. and Rainer H. B. 2011. Application of Gas Chromatography-Mass Spectrometry for Analysis of Isoprostanes: Their Role in Cardiovascular Disease. *Journal of Analytical and Applied Pyrolysis.* 90:13-17.

Hefley, T. J. 1987. Utilization of FPLC-purified bacterial collagenase for the isolation of cells from bone. *Journal of Bone and Mineral Research.*

https://byjus.com/chemistry/classification-of-chromatography/.

https://chem.pg.edu.pl/documents/175289/4235082/AP-Part3.pdf.

https://teaching.shu.ac.uk/hwb/chemistry/tutorials/chrom/chrom1.htm.

Hubschmann, H. *Handbook of GC/MS: Fundamentals and Applications.* Wiley-VCH Verlag, Germany, 2001.

Jones J. R. *Organic Chemistry,* 2nd ed. New York, NY. W. W. Norton & Company, Inc. 2000.

Karger, B. L., L. R, Synder, C. Horvath. 1973. *An Introduction to Separation Science. 1st Edition,* John Wiley and Sons, New York, USA

Krugers, J. *Instrumentation in Gas Chromatography.* Centrex Publishing Company-Eindhoven, Netherlands, 1968.

Lehman, J. W. *Operational Organic Chemistry,* 3rd ed. Upper Saddle River, NJ. Prentice Hall. 2002.

Madadlou, A., Sullivan, S. O., and Sheehan, D. 2011. Fast Protein Liquid Chromatography. *Methods of Molecular Biology.* 681:439-470.

Mohamed, A. J. 2019. Pharmaceutical Applications of Gas Chromatography. *Journal of Applied Sciences.* 9:9.

Moon, P. G., Hwang, H. H., Boo, Y. C., Kwon, J., Cho, J. Y., and Baek, M. C. 2008. Identification of rat urinary glycoproteome captured by three lectins using gel and LC-based proteomics. *ElectropHoresis* 29: 4324–4331.

Plank, J., Andres, P. R., Krause, I. and Winter, C. 2008. Gram scale separation of casein proteins from whole casein on a Source 30Q anion-exchange resin column utilizing fast protein liquid chromatography (FPLC). *Protein expression and purification.*

Riaz, M., Perveen, R., Javed, M. R., Nadeem, H. and Rashid, M. H. 2007. Kinetic and thermodynamic properties of novel glucoamylase from Humicola sp. *Enzyme Microbiol Technology*. 41: 558–564.

Scott, R. P. W. *Chromatographic Detectors: Design, Function, and Operation*. Marcel Dekker, Inc., USA, 1996.

Skoog, D. A. Holler, F. J. and Crouch, S. R. *Principles of Instrumental Analysis.* Sixth Edition, Thomson Brooks/Cole, USA, 2007.

Skoog, D. A. Holler, F. J., Crouch, S. R. *Principles of Instrumental Analysis,* 6th ed. Belmont, CA. Thomson Higher Education. 2007.

Skoog, D. A., F. J. Holler, and S. R. Crouch. 2007. *Principles of Instrumental Analysis.* 6th Edition. Thomson Publishing USA.

Steven J. L. 2002. Application of gas chromatography in food analysis. *Trends in Analytical Chemistry.* 21:686-697.

Tangvarasittichai, N. V. 1993. Use of Fast Protein Size-exclusion Liquid Chromatography to study the Unfolding of Proteins which Denature through the Molten Globule. *Biochemistry.* 32:13288-13298.

Tangvarasittichai, O., and Jermnim, N. 2009. Comparison of Fast Protein Liquid Chromatography (FPLC) With HPLC, Electrophoresis and Microcolumn Chromatography Techniques for the Diagnosis of Beta-Thalassaemia. *Indian Journal of Medical Research.* 129: 242-248.

Thomas G. R., Jeffrey D. H., John M. M. and Thomas A. S. 2000. Fatal Strychnine Poisoning: Application of Gas Chromatography and Tandem Mass Spectrometry. *Journal of Analytical Toxicology* 24:642-647.

Wade, L. G. *Organic Chemistry,* 6th ed. Upper Saddle River, NJ. Prentice Hall. 2006.

Willard, H., L. L. Merritt, J. R., J. A. Dean, F. A. Seattle, Jr. 1981. *Instrumental Methods of Analysis*, 6th Edition. Wadsworth Publishing Company, Belmont, CA USA.

Chapter 9

Analytical Methods Used in Biotechnology

Muhammad Fakhar-e-Alam[1], Muhammd Asif[4], Malik Saadullah[2,*], Muhammad Rashad[2], Muhammad Hanif[3], Muhammad Yasir Ali[5], Muhammd Uzair[3] and Sania Arif[2]

[1]Department of Physics, Government College University, Faisalabad, Pakistan
[2]Department of Pharmaceutical Chemistry, Government College University, Faisalabad, Pakistan
[3]Faculty of Pharmacy, Bahaudin Zakariya University, Multan, Pakistan
[4]Faculty of Pharmacy, Islamia University of Bahawalpur, Pakistan
[5]Department of Pharmaceutics, Government College University, Faisalabad, Pakistan

Abstract

Knowledge and use of appropriate analytical techniques in biotechnology experiments can save a lot of time by eliminating chances of error and trials. Development in biotechnology such as nano technology for pharmaceutical preparations, finding genetic mutations, genetic engineering, and cell culture and cancer biology has increased the use of analytical instruments. The aim of this chapter is to review various analytical techniques used in biotechnology for getting precise and accurate results.

*Corresponding Author's Email: Maliksaadullah@gcuf.edu.pk.

In: Functional Analytical Techniques in Pharmaceutical Chemistry
Editor: Malik Saadullah
ISBN: 979-8-88697-059-3

Keywords: pH meter, mass spectrometry, isotopic analysis and NMR, FTIR and X-ray spectroscopy, HPLC, separative methods, atomic absorption and emission spectroscopy, calorimetry, surface dynamics SPR

Introduction

Biotechnology is an analytical tool for high throughput strain engineering. There are a variety of analytical techniques that are already available for the complete characterization of proteins: 1- for quantitative determination of protein concentration, UV absorbance analysis is used; 2- the secondary and tertiary structure of protein is assessed by spectroscopic methods (X-ray, NMR, circular dichroism); 3- the "top down" approach in mass spectrometry is suitable for measuring the molecular weight of complete proteins, while the "bottom up" approach elucidates the protein sequence using N- and C-terminal sequencing. Furthermore, different chromatographic separation methods (reversed phase (RP), size exclusion (SEC), ion exchange (IEC), hydrophilic interaction chromatography (HILIC), hydrophobic interaction chromatography (HIC), etc. in 1 or 2 direction mechanisms) and spectrophotometry (gel-electrophoresis, capillary electrophoresis) combined to different analytical methods (DAD, MS, Light scattering) may support a more comprehensive analysis.

pH Meter

Much research has been conducted to determine the precise relationship between salivary genetic variants and dental problems. With so many genetic variants and the complications of determining them, dining pH is sometimes used as a simple orthodontic examination. A study was conducted to analyze the pH of saliva and determine its relevance to the severity of periodontal disease. Many microorganisms, particularly if they can grow over a wide range of pH, tailor gene expression to the pH of their growth environment. The genes which are involved in provision of secreted enzymes are most likely affected by ambient pH such as permeases, and exported metabolites, all of which must function at ambient pH, and probably also those involved in internal pH homeostasis. Fungi is used in the production of enzymes, vitamins, polysaccharides, lipids, glycolipids, polyhydric alcohols and pigments, all of these are industrial processes. These kinds of substances are available on the market, although others may be useful in bioengineering. Bioactive compounds produced by fungi are incredibly essential for health and diet, as well as having a significant effect on the economy. The uses of fungi include biotransformation processes and in various reaction sequences of fermentation. Also in the development of single-isomer substrates, which have become increasingly important in the goods market. The use of microbes and other mushrooms as habitats for genetically modified industry has dramatically boosted the economy for anaerobic bacteria. The titer and yield of microbial processes and discovery of new drugs is increased by the addition of molecular manipulations to mutational techniques. The influence of pH on the relative importance of the two major decomposer groups in soil, bacteria and fungi, was investigated along a continuous soil pH gradient at Hoosfield acid strip at Rothamsted Research in the United Kingdom (UK). This experimental location provides a uniform pH gradient, ranging from pH 8.3 to 4.0, within 180 m in a silty loam soil where barley has been continuously grown for more than 100 years. Below pH 4.5 there was universal inhibition of all microbial variables, probably derived from increased inhibitory effects due to release of free aluminum or decreasing plant productivity. Various studies reported the effect of soil pH on fungal and bacterial growth. Similarly, results of studies by Robson (et al., 1996) suggest that alkaline pH gradients may be integral to hyphal extension in fungi. Scientists are trying to modify pH monitoring methods by various means altering electrodes and the incorporation of sensors. Strickler et al. reported the effect of environmental factors on eDNA. He investigated that temperature, pH, and UV-B have some effect on the degradation of eDNA. Maximum parts of DNA are degraded within 10 days, but some eDNA persisted up to 58 days. In cold, low UV-B, and alkaline conditions, degradation rates were lowest.

The experiment starts with the modest pH meter. For survival, most of the time living organisms require neutral pH. The pH range of blood must be 7.35-7.45 and increasing the pH from this range even 1/10 of unit can prove fatal. To increase the yield of crops, soil pH must be controlled. The natural pH of rain water is 5.6 and it is increased to 4.0-5.0 by the addition of atmospheric pollutants which are harmful for living and non-living things.

In the following situations, the control of pH is critical; the prevention of milk from turning sour and preventing harmful effects of cosmetics on skin and eyes.

The pH must be strictly controlled for food manufacturing because changes in pH may show undesirable effects on flavor, consistency and shelf life of the product.

The tension among metal conductors submerged in a mixture is measured by a pH meter, which is a semiconductor device. One of these electrodes is a calomel electrode and the other a glass electrode. The calomel electrode has a constant electrical potential and is the external reference electrode and the glass electrode is a standard test electrode, its electrical potential depends on pH of the solution. The pH of a solution at 25°C can be calculated by the formula as expressed in Figure 9.1.

$$\text{pH of solution} = \frac{E_{glass} - E_{ref}}{0.0591} \quad \text{at } 25^{\circ}\text{C}$$

Substitute value for E calomel electrode

$$\text{pH} = \frac{E_{glass} - 0.250\ \text{V}}{0.0591} \quad \text{at } 25^{\circ}\text{C}$$

Figure 9.1. Calculation of pH of a solution at 25°C.

The electromotive force (emf) of complete cell:

$E = E_{ref} - E_{glass}$.

Where E_{ref} = potential reference electrode or Calomel electrode which at normal temperature is +250 V. E_{glass} = potential of the test electrode or glass electrode which depends on the solution tested.

Operation of pH Meter

1. Soak new/dry electrodes in water/buffer pH 6-7 overnight or in 0.1 M HCl for 12-24 hours. (Note: always store electrodes in distilled water).
2. Fix a "temperature compensations knob" to the temperature of the solution (Note: This is necessary since the equilibrium constant of a reaction i.e., in acid base titration, does vary with temperature).
3. Remove beaker with water, rinse and mop with tissue paper.

4. Dip electrodes into standard buffer with known pH making sure that the electrodes do not touch the bottom or sides of the beaker.
5. pH meter is adjusted to standard pH by adjusting its knob.
6. Remove standard buffer, rinse these electrodes with distilled water mop with tissue paper and dip in solution whose pH is unknown.

Fungal cultivation is among the most widely used biotechnique in diagnosing, biomedical, and genetic research. *Agar-gel* is the common nutrient medium for culturing. The development of fungus on a petri dish induces a pH transition, which can be used to analyze and determine the production or existence of mushrooms by observing such modifications. The pH sensors that are currently available cannot be used for online pH monitoring in small volumes of agar gel. The ratiometric probe was effectively used for the visualization of pH and acidification in pathogenic *Candida albicans*. The probe has H^+ activated reversible closed-open process of spiroboronate. The probe displayed a precise shift in pH from 7.5 to 6.5 in fungal cells which are treated by amiodarone.

Colorimetric Principles for Real-Time Monitoring of pH

Vinayaka et al. presented a miniaturized sensor which has the capability of detecting pH changes in agar-gel culture medium during fungal growth. For the first time, an integrated approach of impedance and colorimetric principles for real-time monitoring of pH changes (7.6 to 2.0) in agar-gels on-chip is demonstrated.

NMR Spectroscopy

The non-invasive and in-vivo information on the sample is gathered by NMR spectroscopy. It is used for the determination of physical and chemical properties of atoms or molecules of the sample. It provides the information about structure, dynamics, reaction state and chemical environment of molecules. The high resolution three dimensional structure of protein is determined by protein NMR spectra. It can also provide information about nucleic acids such as DNA and RNA.

The complex sample preparation is not required for NMR analysis of extractable metabolites of crude plant extracts, it provides fingerprints of metabolites for their HTS (high throughput screening) without sample preparation.

In biomedicine NMR is used for studies of tissues or cells, for diagnosis of disease and for study on extracts of cultured cells. NMR tools flow NMR and some new probe design, the analysis of complex mixtures, combinatorial-chemistry libraries, bound drugs, unstable compounds, samples very small in size, and heterogeneous samples is done by editing NMR spectra and using mixtures without chromatographic separation. The composition of the flu virus particle transmitter was recently determined, leading to the creation of compounds which inhibit the converter and prevent the virus from infecting the host.

Likewise, the structure of the CMAT receptor (an autoimmune disorder associated with cancer metastatic disease) has been determined, allowing an adversarial form to be designed. By helping learners to analyze how bacterial infection agents act and recognize suspected drug strategies, NMR plays a critical role in the production of medicines and therapies for a variety of illnesses, including HIV, influenza, tuberculosis, and cancer.

NMR gained importance in the field of biotechnology because of its ability to provide information about in vivo assays by noninvasive methods such as metabolic pathways and they can be studied by using isotopes such as ^{31}P ^{13}C. NMR technique gained importance in pathways study of genetically engineered plants. NMR is used to compare metabolic pathways of wild type plants and genetically engineered plants. DNA and protein sequencing techniques are quite different then polysaccharide analysis, and for such analysis NMR and mass spectroscopy is applicable to sequence oligo and poly saccharides. Metabolic NMR imaging technique is applicable for optimization of bioreactors for the study of metabolic pathways of the immobilized cells, transport and diffusion of nutrients, oxygen and water.

UV-Visible Spectroscopy

In biochemical research, it is used for the analysis of proteins, enzymes, nucleic acids, oligonucleotide and for enzymatic activity studies. Various nanoparticles have been prepared by using biotechnological approaches and UV visible spectroscopy is used for characterization of nanoparticles. One example of such a study is antimicrobial activity of silver nanoparticles using cell free lysate of *Bacillus subtilis*. The purity of RNA and DNA was assessed by the absorbance ration at 260 nm and 280 nm (the $A_{260/280}$ ratio). When the spectrophotometric analysis of RNA was done in alkaline solution, the ability of protein contamination detection was significantly improved. When measurements were made in Na_2HPO_4 buffer at a pH of 8.5, UV spectral scans revealed that the 260 nm RNA absorption peak detected in water was transferred by 2 nm to a shorter frequency. The DNA location was determined by UV-visible spectrometer with absorbance (A) at 260nm and concentration was determined by using Beer's Law.

Optical Spectroscopy

Optical spectroscopy has the following advantages: (1) it is measurable, (2) it is quick, (3) it could be used at the bedside, and (4) it has exceptional sensitivity to endogenous genetic markers previously present in the tissue. The 'optical biomarkers' are measured more frequently than conventional imaging methods such as contrast enhanced PET, MRI and CT.

FTIR Spectroscopy

FTIR (Fourier-transform infrared spectroscopy) is a method for obtaining an ultrasonic wave of a concrete, fluid, or gas's absorption and scattering. An IR spectroscopy gathers improved temporal information from a large spectral spectrum at the same time. This gives it a major benefit throughout a dissipate spectrometer, that only calculates frequency across a small spectrum of frequencies at a moment.

The imaging and FTIR spectroscopy is used in medical diagnosis and research because it can distinguish between normal and diseased cells of the body. It can diagnose malignancy of cells and can also differentiate b/w their different levels. FTIR with ATR can be used for the monitoring of biochemical processes held on surfaces. FTIR is extensively used for pharmaceutical analysis by identifying functional groups.

Raman Spectroscopy

Raman spectroscopy is a non-destructive chemical analytical technique which gives detailed information about the chemical structure, polymorphy, phase, crystallinity and molecular interactions. Its principle is based on the light's interaction with the chemical bonds of the material.

In Raman spectroscopy, the molecule scatters incident light which is delivered from a high intensity laser light source. Rayleigh scattering occurs when many of the reflected photons have the same wavelengths (or colour) as the sensing element and hence provides no essential data. The scattering of a small amount of light (typically 0.0000001%) at different wavelengths (or colours), depends on the chemical structure of the analyte. This is known as *Raman Scatter*.

The amplitude and wavelength's location of Raman reflected light are seen in a Raman spectrum, which has a series of spikes. Independent connections such as C-C, C=C, N-O, C-H, and sets of ties such as aromatic ring ventilation mode, polymer chain movements, lattice modes, and so forth relate to a particular molecular restitution.

Atomic Absorption Spectroscopy

The atomic absorption spectroscopy (AAS) is used for the estimation of trace metals of biological media such as blood, hair, tissue, urine, etc. The toxicity of these metals can be evaluated for diagnostic decisions and these decisions decides their daily intake limit.

The intensity of an UV spectrophotometer with a fire burning is usually measured in milliseconds. A flame burner is not used for trace analysis because it decrease sensitivity. A graphite furnace is used in place of flame burner for trace analysis to improve sensitivity by various orders of magnitude (in the parts per billion range (ppb)). The applications of AAS are widely distributed in a range of industries including environmental testing, metal analysis, semiconductor manufacturing, petroleum, chemical production, and pharmaceuticals.

Single Crystal XRD

In single crystal XRD the structural information on crystal cells such as atom locations, bond angles and inter atomic dimensions of the crystal lattice is studied. The technique boosts NMR studies and structure of large molecules such as proteins are determined by this technique.

High-Resolution Transmission Electron Microscopy

To generate an interruption frame, elevated x - ray photoelectron spectroscopy (HRTEM) also uses emitted and dispersed lasers. It's a stage comparison picture that can be as tiny as a quartz crystal structure. In this technique the outgoing modulated electron waves interact at very low angles or interfere with itself during propagation through the objective lens. At a level in the picture plane, all particles emanating from the sample are united.

HRTEM has already been widely used effectively for studying the crystalline phase and lattice flaws in a variety of specialized products at the molecular level. It can be used to characterize point defects, piling faults, dislocations, precipitates grain boundaries, and surface structures.

HRTEM is a high-magnification microscope used to study nanomaterials. High resolution makes it ideal and perfect for imaging materials on the atomic scale.

Mass Spectrometery

Mass spectroscopy is an important analytical technique widely used in addition to gel electrophoresis and chromatography. MS is based on the measuring mass to charge the ratio of charged biological analyte such as peptides. Interpretation of various peaks in mass spectrum tells about m/e value of peptides present in biological samples. Duivelshof et al. in 2019 reported analytical workflow, developed a chromatographic methods which is hyphenated to mass spectroscopy (MS) and used it for determination of protein size, charge, hydrophobic, and hydrophilic variants of daratumumab.

PCR

Polymerase chain reaction (PCR) amplified the DNA fragments several fold and is widely used in genetic analysis by the help of DNA fingerprinting and medical tests. The polymerase chain reaction (PCR) is a straightforward technique for amplifying a DNA sequence in vitro and producing real nucleotides. Gene editing a genetic code into a matrix and trying to replicate it in a life form takes days or even weeks of research, but with amplicons it takes just hours. While most biochemical analyses, such as genomic DNA detection with radionuclides, necessitate a large amount of biomedical content, the PCR method only necessitates a small quantity. The modifications in PCR is done by scientists every year such as multiplier PCR, hemi-nested PCR, anchor-ligated PCR, WEB PCR etc. In comparison to earlier approaches, PCR can obtain extra sensitive sensing and increased rates of modulation of complex molecules in very little duration. Sanger sequencing is still the benchmark for detecting alterations in medical care, but amplification refractory mutation method PCR (ARMS-PCR) has gained popularity because it is highly sensitive have less limitations than Sanger sequencing. The technique used for gene detection is mutation-selected amplification specific system PCR (MASS-PCR) because it is a highly specific and sensitive technique. The accuracy level of MASS-PCR and Sanger sequencing (kappa value=0.929) is higher than the consistency rate of MASS-PCR and ARMS-PCR (kappa value=0.821).

MASS-PCR Assay

MASS-PCR is a qPCR-based assay that detects widespread defects in the EGFR, KRAS, BRAF, HER2, MET exon 14 missing mutant, ALK fusion, and ROS1 combustion genomes using bioinformatics tools and queries. This assay was performed in a StepOne™ PCR System according to the instructions given by manufacturers. The gene, mutations, like ALK fusion and ROS1 fusion genes was detected by one-step method.

ARMS-PCR Assay

The re-testing of DNA samples were done by ARMS-PCR. The same mutations are detected by ARMS-PCR as were detected by MASS-PCR assay. For the uncovering of ALK union and ROS1 synthesis genetic factor, the two-step process was required. Concisely, the extracted RNA from FFPE tissue is produced into complementary DNA (cDNA) according to the manufacturer's instructions. The genotype was than determined by cDNA. The kit was bought from the Beijing ACCB Biotechnology Co., Ltd.

Multiplex PCR Assay

In veterinary medicine, the actual PCR is a universal analysis tool, particularly for microbe related illnesses. However, real-time PCR applications in medical mycology have remained challenging.

Sanger Sequencing

The verification of genotype is performed by Sanger sequencing. To intensify the complex protein sequence, the DNA was castoff as model, then the Sanger method is used for sequencing the PCR products.

ELISA

Enzyme linked immunosorbent assay ELISA involves identification of substances on the basis of antibodies antigen interaction. The antigens from the material are attached to the surface. Then specific antibodies are introduced for binding with antigens. The linking of antigen and enzyme is done by enzyme containing substrate. When the reaction is completed, it produces a colour, detected as an absorbance signal.

The HIV blood serum screening tests, determination of food allergens, hepatitis B markers in serum and home pregnancy tests, are all done by ELISA.

DNA Sequencing

DNA Sequencing is used to determine nucleotide sequences in DNA double strands such as adenine, guanine, thymine, and cytosine.

Chromatography

The biomolecules are separated by chromatography so that's why it plays an important role in biotechnology. The high-performance liquid chromatography (HPLC) was introduced in the late 1960s and it expanded the power of this technique. Over the past two decades, the fundamental aspects of the most commonly used modes of separation (reversed phase, ion exchange, etc,) have remained relatively unchanged. Instead, significant progress has come from improvements in column technology, especially in the development of chromatographic supports designed for biomolecules.

Fluorescence Based Detection in Earlier Chromatographic Techniques

HPLC

The high performance liquid chromatography is based on the principle of column chromatography. In traditional column chromatography, firm breathable specimens with a median diameter varying from 30 to 90 m are stuffed into a tubular glass or plastic pipe. The graduated cylinder is made up of the solid adsorbent stored in the pillar, and the liquid phase is made up of the fluid substance(s), which seep via the section underneath gravitational effect. In HPLC, the mobile phase is the liquid solvents and the column is filled up with solid stationary phase. However, the HPLC column is tightly packed with stationary phase material which has a very small diameter (2-50 µm in average particle size). The small packing material increases the surface area, this increased surface area increases the sensitivity and resolution of the process, so that's why it's called high performance.

The fluid working phase cannot easily switch thru the HPLC system due to the massive resistance because as the column becomes smaller and much more densely compacted. The mobile phase is introduced in the column with high pressure by using a pump to increase the speed and resolution of the separation.

The liquid stage and solid phase are chosen based on the analyte's characteristics. The magnetic field of the mobile phase and the stationary phase determines the form of HPLC: regular step or backward state.

Reverse Phase

The reverse phase chromatography is the reverse of the normal phase as it has the polar mobile phase and non-polar stationary phase. For example, when a long hydrophobic C-chain is linked with silica particles (C-8 and C-18 columns) the stationary phase become non-polar. Commonly used mobile phases in reverse phase chromatography are acetonitrile, methanol and tetrahydrofuran (THF).

Normal Phase

In the normal phase HPLC, the mobile phase is hydrophobic or non-polar while the stationary phase is hydrophilic or polar. The stationary phase is composed of water loving substances such as silicon dioxide or materials containing polar groups such as cyano or amino groups. Hexane and methylene chloride are two popular nonpolar reagents used in HPLC. The hydrophobic moving step causes the hydrophobic (non-polar) substances to denature immediately, followed by the hydrophilic (polar) compounds.

Analytical HPLC

The metabolites are isolated in this form for analysis method, such as recognition or quantitative analysis. Analytical HPLC is typically performed in thinner lines, such as those measuring 250 mm long by 4.6 mm in thickness. The flow rate is held between 0.5 and 1 mL/min.

Preparative HPLC

Sample preparation chromatography is the process of separating and isolating multiple components of the sample. The separated items could be used in certain laboratory and field experiments. Therefore, this is used as preparatory step for next experimentation. The particle sizes of the stationary phase in preparative HPLC are bigger in size (10 microns in some cases). Columns usually have an internal surface of 50 to 200 mm and a length of 250 to 1000 mm. The dissolution medium flow rate is held aloft, about 4-5 mL/min.

Instrumentation

The HPLC system consists of solvent reservoirs, pumps for creating high pressure, sample injectors (manual or automated injectors), column, detectors, outlets for waste or fraction collector and a computer system for recording the data.

Solvent Reservoir

The mobile phase or solvent is stored or placed in a solvent reservoir. It may contain a single solvent or a mixture of solvents. Some instruments can introduce up to 4 solvents. When using a regression form, offering several liquid reserve choices is advantageous. Until injecting the resin into the system, it must be equilibrated to eliminate any air bubbles that might obstruct the movement of the extraction solvent via the section. The mobile phase should be degassed by some instrument or manually before its introduction in the column.

There are two types of elution that are mostly used 1- *isocratic elution,* in which mobile phase composition is same throughout the experiment. 2- *Gradient elution,* refers to a mechanism in which the mobile step is shifted at various times. It is reversed - phase elution, for example, if the mobile step is 10% methanol and 90% liquid during the race. If the solvents are changed from 10% methanol (90% water) to 3% methanol (70% water) to 50% methanol (50% water) at the end, the elution is called *gradient elution.*

Pump

The column of HPLC is small in size and densely populated. Force prevents the cellular process from passing around. The pump is placed between the solvent reservoir and column. The pump continuously transfers mobile phase to the column at constant pressure and volume. The peristaltic pump is the pump which is commonly used in HPLC.

Sample Injector

Mostly, the automated sample injector is fixed with the modern instruments and it injects sample from sample tube and this sample is injected to HPLC column, then this sample is carried out with mobile phase in column.

Column

The HPLC columns, as stated earlier, are densely packed with tiny spheres with a size distribution of 2–50 m. Silica or hydroxyapatite materials, as well as polyethylene adhesives like polystyrene divinylbenzene, are popular wrapping substances in HPLC columns.

The column material is selected on the basis of sample properties. Magnetic field, ionic exchange, ionic interference, charge transfer, and other attributes could be manipulated. Different sizes of HPLC columns are described in Figure 9.2.

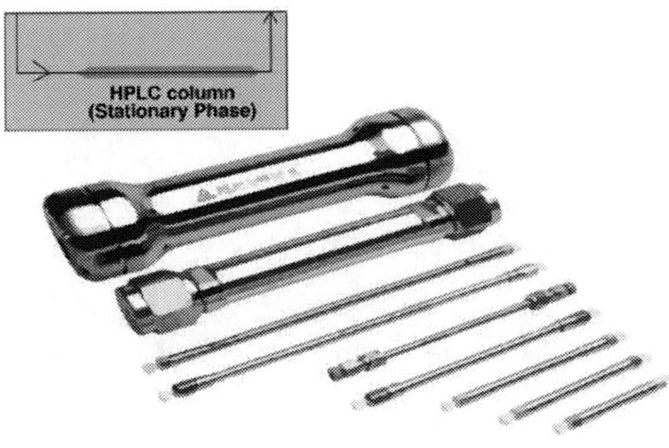

Figure 9.2. Different options in sizes in HPLC columns (Orochem Technologies).

The HPLC column used in range of size, may be analytical or preparative, depends on the purpose of study. The internal diameter (ID) and density of a pole define the amount of sample that must be mounted onto it along with the isolation tolerance. The smaller the column's inner diameter, best is smaller separations and have high sensitivity.

There is a column which is similar to HPLC column but smaller in size, which is placed in between sample injector and the analytical column, known as *guard column.* It protects the main column from contaminants, if any, in the sample. It increases the lifespan and resolution of analytical column.

Detector

As the sample passes through stationary phase with mobile phase, the components of sample are separated. Then the separated components pass through a detector, this detector detects the existence of any complex accepted by the mobile phase. The detectors used in HPLC are discussed below.

UV, VIS, and PDA Detectors

Mobile phase portion. The sample's amount is determined by the measurement of difference between absorption of eluent flowing through column with and without the sample. These detectors can detect the substance at pg level.

The UV sensor is a popular HPLC tracker. The spectrum of a UV detector will range from 195 to 370 nm, while the VIS device requires a different range (400 to 700 nm).

PDA detectors, also described as a UV detector, detectors (DAD detectors), monitor a beam of colours rather than a specific light. The illumination from the lasers is directed into the liquid sample in this process. The photon is then scattered using a laser source, and the intensity of dispersed light in the phototransistor clusters is measured for various wavelengths. Hence a complete spectrum is achieved. The difference in absorbed wavelength or the difference in spectrum is detected in the presence and absence of analytes. A PDA detector is illustrated in Figure 9.3.

Reflex Index Detector

The difference in reflex index within the sample cell carrying the sewage and the reference cell comprising only the solvent system is measured by the RI detectors. The reflex index and angle of the incident light is changed when any compound is present in the effluent. This change can be measured and detected. Nonetheless RI gauge has minor compassion with respect to UV detector. It is useful for the detection of those compounds which do not absorb UV, such as alcohol, sugar or inorganic ions.

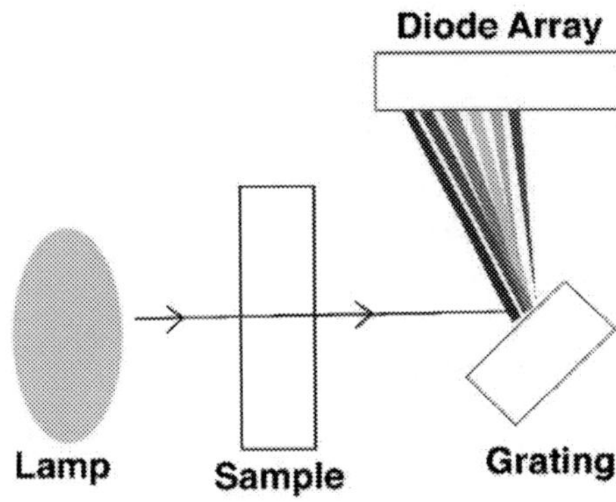

Figure 9.3. Representation of PDA detector.

Evaporative Light Scattering Detector (ELSD)

An ELS detector detects those compounds which are non-volatile or semi-volatile with good and greater sensitivity (at ng level). A nasal inhaler converts the sewage into the spray nozzle, and the liquid phase is warmed and dissipated. The diffraction pattern is observed after the illumination (laser beam) passes through the disintegrated material. Sugar, lipids and high molecular weight analytes are detected by this method.

Mass Spectrometer

For the purpose of further detection and analysis, the HPLC can be connected to a mass spectrometer called as LC-MS. The analytes can be identified by their masses. If the composition of the reagents is unclear, the spectrophotometer will classify it.

Waste Collector or/and Fraction Collector

The effluent is collected in preparative HPLC and discarded in analytical HPLC after its flow through detector. The fraction collector connected to detector for taking effluents from detector for further analysis and studies if any compound is present in the analyte.

Computation and Data Analysis

Eventually, the sensors are linked to a device, which monitors the information in the form of a chromatography paper. The chromatogram provides different data like retention time and amount of sample components.

Size Exclusion Chromatography

SEC, also known as molecular sieve chromatography, is a spectrophotometric process that separates metabolites in solutions based on their size and, in certain situations, molar mass. Complex particles or biological macromolecules, like proteins and cognitive structures, are commonly used. Gel-filtration gel electrophoresis is used where an aqueous solution is used to move the specimen via the device, as opposed to high performance liquid chromatography, that is used whenever a liquid product has been used as a moving step. The stationary phase filled in column is mostly fine, porous beads which are composed of agarose (Sepharose), dextran polymers (Sephadex), or polyacrylamide (Sephacryl or BioGel P). For the estimation of the dimensions of macromolecules, the pore sizes of these beads are used. SEC is widely used in the characterization of polymers because of its ability to provide good molar mass distribution (Mw) results for polymers.

Ion Exchange Chromatography

The ion-exchange chromatography (IEC) is an analytical technique used for the separation and determination of ionic compounds. These compounds are separated on the basis of ion-partition/interaction and ion-exclusion chromatography. The principle of ion exchange chromatography includes, separation on the basis of ionic interaction between ionic and polar analytes, in which ions are present in eluent and ionic functional groups fixed to the chromatographic support. There are two distinct mechanisms which are followed by ion chromatography. One in which the ion exchange is due to competitive ionic binding (attraction) and second in which ion exclusion is due to repulsion between similarly charged analyte ions and the ions fixed on the chromatographic support, play an important role in the separation. The above mass spectroscopy is among the most important adsorption strategies used throughout the segregation of charge holding entities of diverse biological dimensions and molecular composition, such as peptides, proteins, nucleic acids and associated biomaterials. The separation is based upon the formation of ionic bonds between the charged groups of biomolecules and an ion-exchange gel/support carrying the opposite charges. Because of certain various charge characteristics, organic molecules associate with activated spectroscopy medium to various extents. The ion exchange chromatographic method was first reported by Peterson and Sober in 1956, for the separation of proteins, then in 1975 Stevens and Bauman introduced the modern form of ion exchange chromatography. In 1979, Gjerde et al. published an oxidizing agent mass spectrometry process, which was preceded by an acceptor spectrometry technique in 1980. X-ray crystallography is often used to distinguish various organic species, including analytes, for several decades and is being used nowadays. The popularity of this method has been increased in recent years because it allows the analysis of a wide range of molecules in pharmaceutical, biotechnology, agriculture, environmental and other industries.

Affinity Chromatography

Affinity chromatography is a technique for extracting biomolecules from mixtures that rely on an extremely specialized supramolecular covalent attachment within the biomolecule and further material.

Affinity chromatography is a technique for isolating a goal analyte from a strange combination. It can be used to detoxify enzymes, sugars, medicines, haptens, or any other electrode surface of significance when an association couple has been identified. This technique involves manipulation of a specific type of interactions between the binding affinity pairs, like the interaction of antibody to its specific antigen and interaction of ligand with its specific receptor/protein. In 1970 the immuno-affinity chromatography was introduced for the discovery of protein G and L, after that this technique gained popularity as standard methodology for the purification of antibodies and now its applications have become broadened such as being used for immunodiagnostics.

TLC

There is a process called TLC-Bioautography, used for the assessment of inhibitory effect of those compounds which are present in plant extracts against microbes. However, this method has a hidden, currently underutilized potential for detecting and evaluating the existence of inhibitory compounds against selected enzymes. The aim of this process is to evaluate the presence of protease inhibitors in plant extracts. The method is based on the hydrolysis of *Nα*-benzoyl-DL-arginine-*p*-nitroanilide hydrochloride (BA*p*NA) by *α*-chymotrypsin as a representative serine protease to produce coloured *para*-nitroaniline (*p*NA). The production of a pink azo dye is achieved by derivatizing pNA with sodium nitrite and N-(1-napHthyl) ethylenediamine (NPED). This phase increases the chromatogram's clarity of chemical ingredients that behave as bright patches against a violet backdrop. This method was used for the analysis of protease inhibitors in different plant parts such as grape pomace from *Vitis vinifera*, *Picea abies* bark, *HippopHae rhammoides* berries, *Triticum aestivum* bran, *Hordeum sativum* bran and *Avena sativa* bran. The whole approach was utilized to evaluate leaf extract that cannot be examined by a widely established UV spectrophotometer owing to intervention. Those plant extracts are assessed by this method which cannot be analyzed by common spectrophotometric method due to interference.

Primary Amino Acid Sequence of Proteins

In protein structures, the amino acids are arranged in linear sequence, making peptide chains that convert to proteins. The basic amino acid sequence starts from amino-terminal (N) end to the carboxyl-terminal (C) end, as is customary. The ribosomes in cells are used for protein biosynthesis.

Peptide Mapping

The mixture of reduced and denatured proteins is separated in a first dimension by sodium dodecyl sulfate-polyacrylamide slab gel electrophoresis. Following electron microscopy, the entire solution channel is adjusted in mounting gel buffers and moved to a different slate gel at the perpendicular. The solution of protease is overlayed on the gel lane and a partial proteolysis of the proteins to be analyzed is performed during the stacking phase of the second electrophoresis. The other electrophoresis solves the peptide sequence of every other single polymer as a sequence of patches positioned just under the indigestible chromatin current location. The peptide maps of the following samples are shown as examples: protein P23 and P23* of bacteriopHage T4, membranes of Dictyostelium discoideum, membranes of human erythrocytes, and 35S-labeled proteins of D. discoideum manufactured in vivo or in a cell-free wheat germ excerpt. Up to 20 independent proteins could be studied all together in biological mixtures, and an antibody accounting for just 1% of the full number produces an immediately recognizable protease template. Good reproducibility of the patterns obtained allows the comparison of samples of various origins.

Glycoproteins

The proteins which comprise oligosaccharide (glycans) chain covalently attached with amino acid side-chain are called glycoproteins. The attachment of carbohydrates to protein in posttranslational modification and it is known as *glycosylation*. Mostly the secreted extracellular proteins are glycosylated. The glycoprotein is formed by the attachment of carbohydrates to some proteins.

Extracellular segments are often glycosylated in proteins that have extracellular extending segments. Glycoproteins are essential membrane proteins that play an essential role in cell-cell interaction. It differentiates endoplasmic reticulum- based glycosylation of the secretory system from reversible cytosolic-nuclear glycosylation. The nucleus and cytosol's glycoproteins can be modified by reversible addition of a single GlcNAc residue, which is considered opposite to phosphorylation and its function is to regulate the mechanism that control phosphorylation-based signaling. In contrast, the classical secretory glycosylation is structurally essential. Inhibition of asparagine-linked, i.e., N-linked, glycosylation, for example, can inhibit adequate protein kinase binding and can even be harmful to a single cell. Disturbance of glycoproteins production (oxidative withdrawal of glucose compounds to the glycolipids), which is what arises both within the extracellular matrix and the Golgi complex, was never needed for cells cultured (as evidenced by persistence of sterols antagonists), and it can cause infectious diseases (chronic abnormalities of glycation) and is fatal in animal studies. For endogenous functionality, the suitable proceeding of glycan's is necessary such as cell trafficking, but this is likely to have been secondary to its role in host-pathogen interactions. The ABO blood group network is a well-known illustration of such a former influence.

Gas chromatography (GC) is an experimental method for removing chemical composition in a test solution and afterwards detecting these to ascertain their truth or falsity, as well as how many of each element is available. Organic compounds or bubbles seem to be the most common organic elements. Such molecules must be flexible, with a molar mass beneath 1250 Da, and environmentally benign so that they would not decay in order for GC to be effective in their study.

Gas Chromatography

Gas chromatography (GC) is an experimental method for removing chemical composition in a test solution and afterwards detecting these to ascertain their truth or falsity, as well as how many of each element is available. Organic compounds or bubbles seem to be the most common organic elements. Such molecules must be flexible, with a molar mass beneath 1250 Da, and environmentally benign so that they would not decay, in order for GC to be effective in their study. The GC is used for the quality control of many products and for research purposes as for analysis of natural products and in many industries for the analysis of various toxic substances in food, and in forensic sciences for various investigational procedures. To allow the detection of chemical composition, gas strategic communications are often italicized to mass spectrometers (GC-MS).

Gel Electrophoresis

Electron microscopy is a collection of methods for separating DNA, RNA, and protein molecules based on its size or electromotive force. Gel electrophoresis is used in DNA fingerprinting, detection and purification of nucleic acid and proteins, and for the detection of variants and proteins involved in health and disease. It is often utilized to help identify microbes (infection producing organisms) in plasma or other fluids, as well as in resources like nutrition. Nucleotides or proteins found and processed by agarose gels are often explored using certain DNA sequencing or electron microscopy.

Elemental Analysis in Biotechnology

This is the inductively plasma MS (mass spectrometry), it shows following features: 1- it is highly sensitive and selective for the state of art instrumentation, 2- accurate absolute quantification is possible even in complex matrices, 3- the combination of elemental detectors with chromatographic separation methods make it a versatile technique for later uses, 4- it is the compliment of organic and inorganic MS, 5- for analysis of multi-elements and isotopes.

Conclusion

Chemical compounds are detected, identified, purified, characterized, and quantified using analytical chemistry methods. In biological experimentation these methods are used for research, development and QC (quality control) of pharmaceutical products. In these methods, typically sophisticated instrumentation is used which permit quantitative or qualitative, more or less invasive and destructive analysis of samples. The pH meter, mass spectrometry, isotopic analysis and NMR, FTIR and X-ray spectroscopy, HPLC chromatography, separative methods, atomic absorption and emission, calorimetry, surface dynamics SPR techniques are used for the analysis of mass, chemical composition, molecular structure, radioactivity, interactions between molecules. The applications of analytical chemistry methods range from the identification of molecules in a sample to the validation of production methods.

Analytical chemistry is inextricably linked to advanced biotechnology. Analytical chemistry methods are needed not only to confirm the identities and purity of recombinant DNA (rDNA) technology products, but also to detect and quantify these substances, their decomposition products, and their metabolites (when used as pharmacons) at trace concentrations.

Conflict of Interest

None to Declare

Consent for Publication

None to Declare

Acknowledgement

None to Declare

References

Betnér I., and Földi, P. 1988. The FMOC-ADAM approach to amino acid analysis. *LC-GC* 6 (9): 832–840.

Bhikhabhai, R., Lindblom, H., Källman, I., and Fägerstam, L. 1989. Automated 2-D peptide mapping: Gel filtration and reversed-phase chromatography. *Am. Lab.* May: 76–81.

Burlingame, A.L., Maltby, D., Russell, D.H., and Holland, P.T. 1988. Mass spectrometry. *Anal. Chem.* 60: 294R - 342R.

Christian, G.D., and O'Reilly, J.E. 1986. *Instrumental Analysis,* 2nd. ed. Boston: Allyn and Bacon.

Croft, L.R. 1980. *Introduction to Protein Sequence Analysis.* New York: Wiley.

Danielson, N.D., and Kirkland, J.J. 1987. Synthesis and characterization of 2μm wide-pore silica microspheres as column packing for the reversed-phase liquid chromatography of peptides and proteins. *Anal. Chem.* 59: 2501–2506.

Edman, P. 1950. Method for the determination of the amino acid sequence in peptides. *Acta Chem. Scand.* 4: 283–293.

Edman, P., and Begg, G. 1967. A protein sequenator. *Eur. J. Biochem.* 1: 80–91.

El Rassi, Z., and Horvath, C. 1986. Hydrophobic interaction chromatography of t-RNA's and proteins. *J. Liquid Chromatogr.* 9 (15): 3245–3268.

Fontana, A., and Gross, E. 1986. Fragmentation of peptides by chemical methods. In *Practical Protein Chemistry: A Handbook,* Pp. 67–120. A. Dabre, ed. New York: Wiley.

Garnick, R.L., Ross, M.J., and du Mée, C.P. 1988. Analysis of recombinant biologicals. In *Encyclopedia of Pharmaceutical Technology,* Vol. 1. J. Swarbrick and J.C. Boylan, eds. New York: Dekker.

Glad, M., Ohlson, S., Hansson, L., Mänsson, M.-O., and Mosbach, K. 1980. High-performance liquid affinity chromatography of nucleosides, nucleotides and carbohydrates with boronic acid-substituted microparticulate silica. *J. Chromatogr.* 200: 254–260.

Hartman, P.A., Stodola, J.D., Harbour, G.C., and Hoogerheide, J.G. 1986. Reversed phase high performance liquid chromatography mapping of bovine somatotropin. *J. Chromatogr.* 360: 385–395.

Hayashi, R. 1976. *Carboxypeptidase Y. Methods Enzymol.* XLV: 568–587.

Hjertén, S., and Yang, D.J. 1984. High-performance liquid chromatographic separations on dihydroxyboryl-agarose. *Chromatography* 316: 301–309.

Hunkapiller, M.W., Hewick, R.M., Dreyer, W.J., and Hood, L.E. 1983. High sensitivity sequencing with a gas phase sequenator. *Methods Enzymol.* 91(Part I):399–413.

Kaye, W., and Havlik, A.J. 1973. Low angle laser light scattering-absolute calibration. *Appl. Opt.* 12: 541–550.

Kenley, R.A., Hamme, K.J., Lee, M.O., and Tom, J. 1987. Silica-based size exclusion chromatography to characterize the decapeptide nafarelin in a controlled release pharmaceutical formulation. *Anal. Chem.* 59: 2050–2054.

Larsson, P.-O., Glad, M., Hansson, L., Mânsson, M.-O., Ohlson, S., and Mosbach, K. 1983. High-performance liquid affinity chromatography. In *Advances in Chroma-tography,* Vol. 21, ch. 2. J.C. Giddings, E. Grushka, J. Cazes, and P.R. Brown, eds. New York: Dekker.

Ogden, G., and Földi, P. 1987. Amino Acid Analysis: An overview of current methods. *LC-GC* 5 (1): 28–40.

Pilosof, D., Kim, H.-Y., Vestal, M.L., and Dyckes, D.F. 1984. Direct monitoring of sequential enzymatic hydrolysis of peptides by thermospray mass spectrometry. *Biomed. Mass Spectrom.* 11: 403–407.

Regnier, F. 1987. Liquid chromatography as a tool for differentiating among similar structural forms of proteins. *LC-GC* 5 (1): 230–234.

Rubenstein, M., Rubenstein, S., Familletti, P.C., Miller, R.S., Waldman, A.A., and Pestka, S. 1979. Human leucocyte interferon: Production-, purification to homgeneity and initial characterization. *Proc. Natl. Acad. Sci.* U.S.A. 76: 640–644.

Schlesinger, D.H., Weiss, J., and Audhya, T.K. 1979. Isocratic resolution of amino acid thiohydantoins by high performance liquid chromatography. *Anal. Biochem.* 95: 494–496.

Shively, J.E. 1989. Protein chemistry in the 1990's: Part 1. *Pharm. Technol.* February: 32–44.

Skoog, D.A. 1987. *Principles of Instrumental Analysis,* 2nd ed. Philadelphia: Saunders.

Snyder, L.R., and Kirkland, J.J. 1979. *Introduction to Modern Liquid Chromatography*, 2nd ed. New York: Wiley-Interscience.

Stark, G.R. 1972. Sequential degradation of peptides and proteins from their COOH termini with ammonium thiocyanate and acetic anhydride. *Methods Enzymol. XXV:* 369–384.

Stuting, H.H., Krull, I.S., Mhatre, R., Krzysko, S.C., and Barth, H.G. 1989. High performance liquid chromatography of biopolymers using on-line laser light scattering photometry. *LC-GC.* 7 (5): 402–417.

Tsugita, A., Ataka, T., and Uchida, T. 1987. Approaches for submicrosequencing. *J. Prot. Chem.* 6: 121–130.

Wehr, C.T., Lundgard, R.P., and Nugent, K.D. 1989. Hydrophobic proteins: A challenge for reversed-phase HPLC. *LC-GC.* 7 (1): 32–37.

Willard, H.H., Merritt, L.L., Dean, J.A., and Settle, F.A., Jr. 1988. *Instrumental Methods of Analysis*, 7th ed. Belmont, Calif: Wadsworth.

Willard, H.H., Merritt, L.L., Dean, J.A., and Settle, F.A., Jr. 1988. *Instrumental Methods of Analysis,* 7th ed., Chaps. 19, 20. Belmont, Calif.: Wadsworth.

Wyatt, P.J., Jackson, C., and Wyatt, G.K. 1988. Absolute GPC determinations of molecular weights and sizes from light scattering. *Am. Lab.* May: 86–91.

About the Editor

Malik Saadullah
Government College University Faisalabad
Faisalabad Punjab, Pakistan 38000

List of Contributors

Muhammad Rashad	Department of Pharmaceutical Chemistry, Government College University Faisalabad, Faisalabad, Pakistan.
Ruby Bashir	Department of Pharmaceutical Chemistry, Government College University Faisalabad, Faisalabad, Pakistan.
Jahan Ara Batool	Department of Pharmaceutical Chemistry, Government College University Faisalabad, Faisalabad, Pakistan.
Sana Mahmood	Department of Pharmaceutical Chemistry, Government College University Faisalabad, Faisalabad, Pakistan.
Akhtar Rasool	Faculty of Pharmacy, Government College University Faisalabad, Faisalabad, Pakistan.
Shahid Shah	Faculty of Pharmacy, Government College University Faisalabad, Faisalabad, Pakistan.
Muhammad Ajmal Shah	Faculty of Pharmacy, Government College University Faisalabad, Faisalabad, Pakistan.
Muhammad Fakhar-e-Alam	Department of Physics, Government College University Faisalabad, Faisalabad, Pakistan.
Muhammad Asif	Department of Pharmacology, Faculty of Pharmacy, the Islamia University of Bahawalpur, Bahawalpur, Pakistan
Saeed Ahmad Rao	Department of Pharmacology, Faculty of Pharmacy, the Islamia University of Bahawalpur, Bahawalpur, Pakistan

Muhammad Uzair Faculty of Pharmacy, Bahauddin Zakariya University Multan, Multan, Pakistan.

Bashir Ahmad Ch. Faculty of Pharmacy, Bahauddin Zakariya University Multan, Multan, Pakistan.

Samina Afzal Faculty of Pharmacy, Bahauddin Zakariya University Multan, Multan, Pakistan.

Muhammad Hanif Faculty of Pharmacy, Bahauddin Zakariya University Multan, Multan, Pakistan.

Durr-e-Shahwar Malik Peoples University of Medical and Health Sciences, Nawabshah, Sindh, Pakistan

Ghulam Abbas Faculty of Pharmacy, Government College University Faisalabad, Faisalabad, Pakistan.

Sania Arif Department of Pharmaceutical Chemistry, Government College University Faisalabad, Faisalabad, Pakistan.

Bisma Kanwal Department of Pharmaceutical Chemistry, Government College University Faisalabad, Faisalabad, Pakistan.

Index

D

E

F

G

H

I

L

M

N

O

P

Q

R

S

T

V

X